TAKING CHANCES

TAKING CHANCES

The Coast after Hurricane Sandy

EDITED BY KAREN M. O'NEILL
AND DANIEL J. VAN ABS

RUTGERS UNIVERSITY PRESS
New Brunswick, New Jersey, and London

Library of Congress Cataloging-in-Publication Data
Names: O'Neill, Karen M., 1959– editor. | Van Abs, Daniel J., editor.
Title: Taking chances : the coast after Hurricane Sandy / edited by Karen M. O'Neill and Daniel J. Van Abs.
Description: New Brunswick, New Jersey : Rutgers University Press, [2016] | Includes bibliographical references and index.
Identifiers: LCCN 2015032499| ISBN 9780813573779 (hardcover : alk. paper) | ISBN 9780813573762 (pbk. : alk. paper) | ISBN 9780813573786 (e-book (epub)) | ISBN 9780813573793 (e-book (web pdf))
Subjects: LCSH: Hurricane Sandy, 2012. | Hurricanes—Environmental aspects—Atlantic States. | Coastal zone management—Atlantic Coast (U.S.)
Classification: LCC QH104.5.A84 T35 2016 | DDC 333.9100974—dc23
LC record available at http://lccn.loc.gov/2015032499

A British Cataloging-in-Publication record for this book is available from the British Library.

Visit our website: http://rutgerspress.rutgers.edu

Manufactured in the United States of America

For Bernard O'Neill and in memory of Barbara O'Neill

In memory of Patricia P. Van Abs

CONTENTS

ACKNOWLEDGMENTS

Soon after Hurricane Sandy hit, a group of researchers convened, with the shared interest of documenting reaction to the storm. Geographer Robin Leichenko crystallized the sense of that meeting and proposed what became the main question for this book, whether Sandy would be transformational. We thank Robin for this idea and for providing funding from the Rutgers University Climate Institute to support this book. Funds also came from Karen O'Neill's work as a member of the Sasaki/Rutgers/Arup team in the Rebuild by Design competition. With support from the Rockefeller Foundation and other charitable organizations, Rebuild by Design produced resiliency designs for coastal areas in response to Hurricane Sandy. Daniel Van Abs and Karen also thank the New Jersey Agricultural Experiment Station and Rutgers University's School of Environmental and Biological Sciences, institutions that encourage us to link basic research to practical applications in response to a changing world. Most of all, we thank the wonderful authors of chapters in this book, whose experience and professional expertise can help the rest of us to understand the legacy of this storm.

Research, policy analysis, and design work on coastal resilience in many coastal countries has greatly intensified in recent years in response to hurricanes and typhoons and to new findings about rising sea levels and climate change. Broad networks of policy analysts, data analysts, natural scientists, social scientists, government agencies, and many other institutions have contributed to the questions and new approaches discussed in this book. Most relevant to this book is Dan's extensive work with planning, sustainability, and water resource organizations in New Jersey and Karen's work on Rebuild by Design and on the Baird & Associates team in the Changing Course competition to rebuild Louisiana's coastal lands. Contributors to the book have experiences across many other networks.

Rutgers University Press's commitment to producing books about ecology and about the mid-Atlantic region made it a clear choice as a publisher for this book. Encouragement from Marlie Wasserman, Leslie Mitchner, and Dana Dreibelbis at the Press made this book possible. Two external peer-reviewers provided us the opportunity to reflect back on the significance of events since Hurricane Sandy, giving us helpful comments that affirmed the importance of the questions we raise here. Beth Gianfagna, Marilyn Campbell, and the prepress and promotion staff raised the quality of the manuscript and helped us clarify the message of this book.

Our invaluable colleague Heather Fenyk kept the big picture in mind. Heather has been a long-standing partner of both coeditors in projects ranging from the founding of a new watershed association to researching energy policies. She helped make the chapter drafts into a finished book and provided exceptional research assistance and analysis of proposed and completed efforts to move entire communities away from coastal floodplains and other sites that are being made more hazardous by climate change.

We are especially appreciative of the many individuals who have provided information about their responses to Hurricane Sandy and their reflections on a changing climate. Interviews and surveys done for several of the chapters used methods approved by the institutional review boards of our home institutions. These respondents granted us their informed consent to quote or paraphrase their comments. Other information comes from a wide range of sources, which are credited in the chapters. Images and other figures that are not otherwise attributed were created by chapter authors and are printed here with their permission.

Others who helped along the way are Yveline Alexis, Diane Bates, Mike Brady, Chip Clarke, Cara Cuite, Jill Allen Dixon, Gina Ford, William Hallman, Jason Hellundrung, Brie Hensold, Marjorie Kaplan, Kaysara Khatun, Naa Oyo Kwate, Louis Lemkow, Tania del Mar López-Marrero, Pam McElwee, Mary Nucci, Chris Obropta, Cymie Payne, Rachael Shwom, and Asher Siebert, in addition to those acknowledged within the various chapters. A special thanks goes to Steven Handel and Joanna Burger for sharing their expertise about the interaction of humans with coastal environments. Karen greatly regrets that her colleague and friend Bob Gramling did not live to see revisions of the book after the first draft. She absolves him of any responsibility for mistakes or inelegant writing in the introduction.

TAKING CHANCES

INTRODUCTION

A Transformational Event, Just Another Storm, or Something in Between?

KAREN M. O'NEILL, DANIEL J. VAN ABS, AND ROBERT B. GRAMLING

Hurricane Sandy could change how we live near the sea. People are economically, culturally, and emotionally committed to living and working along the world's shores, but a storm like Sandy reveals how this makes us vulnerable. This enormous storm directly or indirectly killed at least 286 people across the densely settled coastlines of several countries, displaced hundreds of thousands, and caused many billions of dollars in damages. It hit some of the wealthiest areas in the world, as well as some of the poorest. Its path and dynamics were unusual, and even though weather modelers tracked the storm accurately and routinely modified their estimates of its power, they feared that people would underestimate it. Sandy's course from October 22 to 29, 2012, and its remnant effects caused widespread damage across the Caribbean, up the Eastern Seaboard of the United States, and into Canada, unusual for a late-season hurricane.[1] It reduced confidence in Cuba's admired evacuation practices, Manhattan hospitals' expensive power backup systems, and other essential and esteemed institutions. It further damaged the reputations of less-respected institutions across the storm's region, including national disaster agencies, regional transit systems, and electric utilities.

The storm also hit at a time when ordinary people and the media were becoming more willing to link extreme storms to climate change. This trend was especially watched in the United States, where long campaigns by the oil and gas industry and allies had persuaded many to dismiss climate science.[2] Discussion of the United States is central in this book because many have sought its leadership in reducing carbon emissions and because it seemed that Sandy might galvanize new action by that country. Before Sandy, opinion polls in the United States revealed a polarized public but also showed increasing numbers of people who accepted the idea that the climate is changing because of human actions.[3] After the storm, many were willing to directly attribute Hurricane Sandy to climate change (see chapter 4 about post-storm polling). *Bloomberg Businessweek*'s cover after the storm was a nighttime photo of a flooded street in lower Manhattan, illuminated only by a few emergency lights, with the title, "It's Global Warming, Stupid."[4] The editors acknowledged that statements like that, blaming a single storm on climate change, go well beyond climate scientists' interpretations that climate change is making extreme storms more likely, but they wanted to make a point. Politicians in the United States and in many other countries have been even more cautious about making these direct links, although after Sandy, Governor Andrew Cuomo of New York said: "Climate change is a reality, extreme weather is a reality, it is a reality that we are vulnerable. . . . There's only so long you can say, 'this is once in a lifetime and it's not going to happen again.' The frequency is way up. It is not prudent to sit here, I believe, to sit here and say it's not going to happen again. Protecting this state from coastal flooding is a massive, massive undertaking. But it's a conversation I think is overdue."[5]

The authors of this book felt it was urgent to document the period soon after this storm, a time when people might be motivated to make fundamental changes. Is Hurricane Sandy a transformational event, just another storm, or something in between? We ask if Hurricane Sandy has transformed organizations, communities, and policies in the storm region or more broadly, focusing on urbanized areas that increasingly dominate vulnerable coasts around the world.[6] To address this question, we ask if people have changed their perceptions of coastal vulnerability, if they have made new plans in response to Sandy, whether they face constraints to action, what cultural and emotional responses they have had, and what they think should be done over the long run. This book is not about Hurricane Sandy itself, but rather on what policy and operational shifts happened in reaction to the storm.

Robert Kates, William Travis, and Thomas Wilbanks distinguish incremental forms of adaptation to climate change from transformative adaptation. People make incremental adaptations when they extend actions that they are already using, such as building ever higher seawalls to protect a shore town. People make transformative adaptations when they apply a strategy at a larger scale or at a higher intensity, when they create entirely novel approaches or bring a form of adaptation to a particular setting for the first time, or when they utterly transform a place or move entire communities away from harm. Transformation may be led by governments, firms, and nongovernmental organizations, such as the Netherlands' emerging regime that makes "room for the rivers" by allowing some inland flooding. Or transformation may result from the cumulative actions of individuals, such as farmers changing a cultivation practice one by one.[7]

A single disaster rarely results in a transformative adaptation. For people other than the immediate victims, worry usually drops soon after the disaster. But those rare exceptions stand out in our imaginations, giving us the sense that someone will do something to respond to such an obvious problem. The most prominent coastal disaster with such an effect was the 1953 storm surge that killed 1,836 people in the Netherlands, prompting its central government and other institutions to commit to an expensive program of action, including massive sea barriers and changes in land use policies. The Dutch had already built a base of knowledge and engineering expertise, had centuries-old local polder boards that were integral elements of local governance, and had long viewed flooding as an existential threat that required public spending.[8] Long-term institutional and cultural commitments like these make it more likely that people will respond to a new disaster by evaluating what went wrong, exposing long-standing flaws in management systems, and advancing transformative adaptations.

In the United States, some disasters have prompted transformative adaptations to the physical environment by exposing the inadequacy, conflicting duties, or incompetence of parts of the government that are often related to the ambiguities of a federal system. The BP Deepwater Horizon oil spill in 2010 spurred the 2011 reorganization of the former Bureau of Ocean Energy Management, Regulation, and Enforcement to separate decisions about offshore oil leasing from decisions about the standards for such mining.

Regional and state-level initiatives include the creation of the first interstate-federal compact forming the Delaware River Basin Commission in 1961 in response to a 1955 flood, the adoption of statewide building codes in Florida after Hurricane Andrew in 1992, and more stringent seismic requirements for new construction in California after the Northridge earthquake of 1994.[9]

Other efforts seen as transformative at the time created new problems that have inadvertently reduced adaptive capacity. After decades of ad hoc relief efforts after storms, Congress reacted to Hurricane Betsy in 1965 by passing the National Flood Insurance Act of 1968. This program has unintentionally encouraged new development in some hazardous areas by providing coverage that the commercial market was unwilling to provide (because floods typically require payments to many policyholders at once) and by subsidizing premiums for many of the insured.[10] The September 11 terrorist attacks in 2001 provoked a shift of the Federal Emergency Management Administration (FEMA) and other agencies into the new Department of Homeland Security in 2002. Many see the larger agency's security mission as overshadowing FEMA's mission to reduce hazards and respond to disasters.[11]

Smaller changes in policy may prepare the way for greater transformations. The historic Mississippi River floods of 1993 resulted in a shift of thinking by local and federal officials, who used aid money to help relocate to higher ground several entire towns affected by extensive repetitive losses.[12] The failure of New Orleans levees during Hurricane Katrina in 2005 prompted a major reevaluation of the hodgepodge system of navigation planning and levee construction, maintenance, and inspection in the area.[13] In response to Hurricane Sandy, FEMA began requiring states to assess the effects of climate change when writing hazard mitigation plans, documents they must submit to receive disaster preparedness aid.[14]

As these examples show, for events to be influential, they must be interpreted as indicating recognizable broader problems. Whether Hurricane Sandy is transformational in the long run may depend on the nature of the worries that people have after this storm and whether they see it as an anomaly or as part of a trend pushed by climate change. Each of these two interpretations has its intuitive appeal. And even if individuals are convinced that disaster costs are high or that climate change is occurring, they may still make decisions that attempt to sustain the status quo. A politician is reluctant to spend money today to raise roads or bridges that may not be flooded for years. A retiree considers her new oceanfront condominium a great investment and thinks it is unlikely that a storm will force her to evacuate while she lives there. Rationalization may trump rationale.

Yet Hurricane Sandy might keep people's attention. It hit places in the Caribbean and the United States that were thought safe and it swamped parts of a global city that hosts the United Nations, major global media outlets, and major corporations.

This book builds on the cornerstone idea of hazard and disaster studies, that a storm or drought only becomes a disaster because of the many large and small decisions that make us vulnerable.[15] Hazards are created when people expose themselves to potentially harmful conditions. We try to manage those hazards by building shelters and protective infrastructure and imposing or accepting regulations that constrain where we live, how we build, and how we behave. The storm would be transformational if it led to new developments in science and technology and changed institutional processes and social behavior in ways that made us fundamentally less exposed to hazards.

Unhappily, the seas are rising at an accelerated pace, at a time when more people around the world live in hazardous coastal areas than ever.[16] Coasts have historically supported fishing and shipping activities, and port cities have also supported manufacturing, banking, and cultural activities. Over the past one hundred years, increased agricultural productivity, declining rural opportunities, and improved transportation and other technologies have pushed or encouraged millions more to migrate from rural areas to the coasts.[17]

Community ties, place attachment, culture, infrastructure, and money create powerful incentives for residents and businesses to stay in hazardous coastal places and for newcomers to join them.[18] About 10 percent of the world's population and 14 percent of the population in the least developed countries live in a "low elevation coastal zone," less than ten meters above sea level and contiguous with the coastline. The complex social and physical infrastructure of much of this area is vulnerable to storm surge and ocean flooding. Some of the lowest-lying areas around the world have therefore become some of the most highly priced or densely settled, including 65 percent of cities with more than five million residents. This has resulted in growing costs and human losses from nuisance flooding and disasters.[19] Wealthy, middle-class, and poor people are increasing their exposure to coastal hazards in Ho Chi Minh City, Miami, Mumbai, and other cities.[20] Urban forms of development have also spread well beyond the big cities. These range from amenity-rich sites with tourist hotels or second homes, where properties most exposed to the ocean are worth many times more than inland properties,[21] to sites with less picturesque economic activities and more modest housing, where owners, renters, or informal settlers have little economic margin for absorbing the costs of the next storm.[22] Efforts to reduce exposure to hazards would have to work against the many attractions and forms of investment that keep us near the shore.

The United Nations and others have identified protection, accommodation, and retreat as three strategies for responding to climate change. Building expensive hard or soft protections is questionable as a long-term strategy to sea level rise for most sites.[23] Soft engineered solutions like artificial dunes or flat expanses of replenished sand on beach sections are temporary and provide only partial protection against surge, and even less against routine flooding caused by sea level rise. Structural protections or hardscapes such as seawalls, jetties, and bulkheads attempt to fix shifting beaches in place or to harden crumbling bluffs. These structures interrupt flows of sand, sediment, water, and species, providing local protection against small or moderate storms but often creating unintended problems for neighboring sites. Barriers would have to be built ever higher and longer as the seas rise, walling off developed areas from the ocean views that help to make the property valuable. Barrier approaches are also seldom affordable for low-value real estate. And feasible structural solutions have not yet been devised to block the flooding of complex estuaries and the back bays of barrier islands or to prevent seawater intrusion into porous limestone substrates as in Miami and the Yucatán Peninsula.

Many coastal experts see accommodation or managed retreat as methods that will ultimately be less socially disruptive, less costly, and less prone to catastrophic failures than are structural barriers. (Chapter 10 explains the surprising fiscal advantages for local governments in the United States of preventing rebuilding on flood-damaged lots.) Accommodation and managed retreat are easiest to envision for new settlements and for areas that are rebuilding after disasters, although retrofitting is sometimes feasible, especially when structures like roads undergo renovations. New York City's comprehensive review of strategies after Hurricane Sandy concluded that barriers would remain its major strategy, that accommodation could be used in sites near natural areas, and that retreat could be practical in only a few sites.[24] Iconic examples of the accommodation strategy, at two economic extremes, are vernacular stilt houses in the Philippines and Vietnam and vacation houses raised high on pilings on barrier islands. Novel designs include floating buildings or floodable streets. Green infrastructure installations, such as restored mangroves or oyster reefs to block wave action, broaden accommodation strategies across wider areas. Some accommodations may be temporary fixes, if houses are not raised to sufficient heights or if coastal wetlands are overwhelmed by rising sea levels.

Managed retreat, or managed realignment, is the most ambitious and politically difficult strategy. Some retreat projects have ecological goals, allowing natural processes to once again dominate a landscape that has been managed. Examples include allowing a sandy beach to erode or removing flood control structures along an estuary.[25] Other projects aim to reduce human exposure to hazards and can be viewed as a form of migration or "planned population movement" that involves abandoning human structures in highly vulnerable areas and resettling residents.[26] The most ambitious hazard mitigation relocation projects aim to move entire communities as a group, rather than dispersing individual households to various new locations. Group relocations tend to involve small and culturally cohesive groups, such as tribes or villages of indigenous people. The most prominent demands for group relocation in response to climate change have been framed as cases of environmental injustice and have centered on claims against energy companies or governments.[27] To date, few managed retreat projects of any kind have been undertaken.

Managed retreat from the shore is unappealing to many people for many reasons. None of the institutions or organizations studied in this book have taken serious steps to pursue managed retreat on the coastal shore since Hurricane Sandy, though some neighborhoods have been purchased for open space in non-shore areas. Throughout this book, we discuss many reasons why people resist leaving. Most people feel they do not have the resources to move their homes or businesses without assistance. Poor people may find it especially difficult and unappealing to make a planned move that would disrupt family and community networks that materially support their daily lives. Wealthier people may stay because they feel they are able to evacuate and to rebuild their coastal homes as needed. And coastal settlements can flourish through government encouragement or through government neglect. In coastal areas with close government scrutiny, governments may create financial incentives for land development and build roads and bridges, transit networks, and flood protections. In areas with less government scrutiny, settlements may spread in hazardous zones because the government has posed few restrictions and because settlers are priced out of less hazardous areas. In either situation, daily activities and the presence of other residents make these places seem normal. Managed retreat is an option that few of these residents could even imagine, unless a storm created such trauma or financial hardship that they could not rebuild. (Chapter 11 discusses residents in these circumstances in one shore town.)

The alternative to deliberately using the strategies of protection, accommodation, or retreat, alone or in combination, is to allow existing settlement patterns and emerging trends to determine who is exposed to coastal hazards and who pays. This is the current practice in both developed and developing countries. The status quo leaves decisions to retreat or recover largely up to individuals and families under duress, leaves households to suffer largely in isolation, and hides the broader social costs of coastal exposure. In extreme circumstances, poor and middle-income people have been forced into temporary or permanent "distress migration" after natural disasters such as Hurricane Mitch in 1998 in Central America and Hurricane Katrina in 2005 in the U.S. Gulf region.[28] Residents to a certain extent may be able to draw lessons about adaptation from their understandings of past conditions, but climate change will soon defy lessons from the past about tides, storms, ecology, and safety.[29]

At the same time that coastal population has grown, government budget cuts and demands for deregulation in many countries have reduced government's ability to manage coastal complexity, leaving it unclear at times who should coordinate responses to coastal hazards.[30] The insurance industry and other businesses, charitable foundations, and other nongovernmental organizations have created initiatives to reduce risk and increase resilience, but they lack the authority and influence of governments. For managers, researchers, students, and other members of the public, this book will identify the types of institutions at work in Hurricane Sandy's region that are taking action to reduce threats, which in turn could limit future financial, ecological, and human costs.

USING CASES TO UNDERSTAND CHOICES

In presenting case studies, we are searching for lessons about how and why change became possible under some circumstances but not others. We do not aim to comprehensively document damages from or responses to Sandy. Instead we select cases that stand in for broader categories of human experience to gain insights into people's current thinking about hazards. For example, to consider whether a near miss to a catastrophe is enough to encourage people to rebuild differently, one chapter looks at reactions to damage on the island of New Providence in the Bahamas, which many residents had perceived as being protected from hurricanes relative to other Caribbean islands. To identify the steering problems that are arising for highly complex, aging, and vulnerable systems managed by many institutions, another chapter analyzes the electricity sectors in New York and New Jersey. Because coastal change affects such a wide range of human experiences, any selection of cases would necessarily leave out detailed discussions of important sectors and topics. In this book, geomorphic impacts and engineering approaches go largely unaddressed. Other neglected topics include private insurers, hospitals and other health services, mental health services, aid agencies, transit systems, roads, and disaster recovery activities. Still, our case selections convey something of the breadth of experience, ranging from individuals to large institutions and from Haitian tent camps to New York City. This range allows us to address our central question of whether the storm has been transformational.

Because climate change is leading to pervasive but varied effects on life on the urbanized coast, we present case studies from across disciplines and across social sectors. Hurricane Sandy challenged all of us as authors, and for some of us, it challenged us personally. Many contributors to this book are in the doom business, having studied and advised others about hazards, risk perception, and the ways organizations manage disasters and environmental problems. Other authors specialize in key institutions like electric utilities, which are now being required by market and political demands to prepare more deliberately for future storms. All of us have found surprises in this storm and its aftermath. Case studies are especially valuable for this kind of exploratory research of phenomena that are new to the world, where theories and hypotheses based on past experiences may fail to capture what is unfolding.

THE CONTEXT

To understand these case studies, we must identify and understand their context, the experiences and ideas that influenced people responding to the storm, and the existing infrastructure, policies, and other conditions that affected how they reacted. The storm was unusual in its track and formation, which worried forecasters and caused difficulties for emergency managers deciding whether to order evacuations. Most people think about the last storm they experienced when they consider whether to evacuate. In the Caribbean and in the United States, comparisons varied across island groups and states, depending on which recent storms they had experienced. People in some hurricane-prone areas, like Puerto Rico and Florida, watched the storm track safely away from their shores. Other areas unused to hurricanes, such as Cuba's southeastern shores and inland areas of Connecticut, were hit hard. Because the storm track was unusual as it developed, in some places people who decided not to evacuate suffered, emergency procedures had to be rushed into action, and emergency equipment was left vulnerable or poorly positioned to respond.

The storm was reported around the world, and even though there were at least 120 deaths in the Caribbean, reporting focused on the United States. In Haiti, which suffered the most deaths, at least 104, people in rural settlements had to recover largely without the help of government or nongovernmental organizations. Most of the agencies and organizations that were already in the field in Haiti were absorbed with managing the misery caused by Sandy for the hundreds of thousands of people who were still displaced by the massive earthquake in 2010, and with the additional thousands whom the storm newly displaced. In Cuba, officials and citizens had to recover from devastation along a coastal section that rarely had hurricanes (see chapter 3 for an overview of responses in the Caribbean).

After the storm, the most common comparison in the United States was to Hurricane Katrina, even though the two storms were very different. Katrina hit portions of the coastal United States along the Gulf of Mexico in 2005, killing more than 1,800 people, devastating the Mississippi and Alabama coasts and New Orleans, and displacing a quarter of that city's residents. New Orleans nine years later still has a lower population than it did before Katrina, and some of the other coastal Gulf towns have also not yet fully rebuilt.[31] Authors in this book who analyze cases in the United States found a number of changes made in response to Katrina, from evacuation practices to flood insurance rates, which demonstrate the potential influence of a single storm. Ultimately, though, reactions to Sandy were very different than reactions to Katrina.

Katrina prompted rounds of blame for human actions or inactions. Reporting outside that storm's region included video of cars evacuating, waters rising, and people on their roofs in New Orleans. As the days passed, the media reported on the many ways that humans have manipulated the Mississippi River, Lake Pontchartrain, and the coastal wetlands, making the city and other Gulf communities more vulnerable. Some commented that people shouldn't live in such a place and asked why residents didn't know enough to evacuate, not understanding that the area had lost protected wetlands owing to flood control systems and oil and gas pipelines and canals, that the city had subsided over the centuries, and that thousands of residents owned no cars.[32] National media coverage of storm damage to areas beyond New Orleans was limited, even though some rural Gulf communities were leveled. Fortunately, evacuation of rural areas was very effective. New Orleans's evacuation processes also worked well for those with cars, primarily because officials converted inbound lanes to the outbound direction. But talk about blame was unavoidable, given the horror of the many deaths caused by Katrina in New Orleans, the special plight of that city's poorest people, and incompetent government efforts to evacuate carless people and to care for the displaced.[33] A few environmentalists and scientists talked about the need for climate adaptation, but calls for action after Katrina focused on reforming New Orleans politics and federal disaster policies and improving preparations for evacuation.[34] Over time, a more technical story also developed about the uncoordinated creation and management of the levee systems surrounding much of the city.

By contrast, when Sandy hit seven years later, it prompted questions almost immediately of whether it and other recent storms were affected by climate change. In the time since Katrina, there had been numerous extreme storms in Asia and the Atlantic and reports about shrinking glaciers and ice sheets. And though Sandy's death toll was considerable, news reports were not dominated by the tone of dread that dominated reporting on Katrina or on Typhoon Haiyan that killed more than five thousand people in the Philippines in 2013.[35]

Instead, early images in the international media of damage from Hurricane Sandy were from lower Manhattan, a place perceived as quite capable of responding to disasters and of reorganizing itself (with substantial financial help from the national government), as it had been doing since the September 11, 2001, terrorist attacks. Destruction to tourist sites like beach boardwalks along the shores of nearby New Jersey and Long Island also attracted media attention, especially when President Barack Obama visited the area.

Sandy stories of hardships for the poor and for ethnic minorities got little attention outside of the region. Deaths from Sandy were most concentrated in Haiti, but media coverage about Haiti treated storm damage as an extension of its previous disasters and ongoing governance troubles. Cuba and other Caribbean islands received even less international coverage. Sandy caused seemingly avoidable deaths in the Staten Island section of New York, in neighborhoods where people in modest beachfront houses had not evacuated. Appalling conditions had also developed after the storm in some New York City public housing high-rise buildings, as a result of underfunding, poor maintenance, and inadequate official relief efforts. Sandy was not associated with obvious racial and class inequality and suffering, as Katrina had been.

Some institutions were prepared to talk about longer-term adaptation. The most immediate and ambitious political action for adaptation was in New York City under Mayor Michael Bloomberg, whose agencies produced the technocrat's dream response. Having already created an ambitious urban plan called PlaNYC, which included a climate change action plan developed in consultation with climate scientists, engineers, designers, and planners,[36] agencies wrote an extensive and expensive resilience action plan after Sandy to improve the city's ability to maintain services and recover from natural disasters.[37] Community groups representing communities with high concentrations of poor and minority ethnic groups called for communities to be involved in the planning directly and for more attention to public health concerns.[38] Mayor Bill de Blasio came into office after Bloomberg, promising more attention to inequality but less discussion of the climate, although he retained Bloomberg's resilience officer and publicly promoted continuation of the PlaNYC work, renaming it OneNYC.[39] As part of Governor Andrew Cuomo's broader New York Rising proposals, the state government led communities throughout the state in planning for climate adaptation and storm preparedness and started measures such as ensuring gasoline supplies after disasters.[40] New Jersey State under Chris Christie worked to coordinate federal aid and speed regulatory approvals for rebuilding, without a focusing planning document.[41] Comprehensive plans began soon after the storm in the Family Islands, a largely rural portion of the Bahamas where people aimed to move settlements away from the shore, to higher elevations (discussed in chapter 3).

As a highly complex set of organizations and practices, the U.S. federal government had programs that were not coordinated. Responses of the National Weather Service and the Department of Homeland Security's role in supporting local evacuation are described in chapters 1 and 2. Regarding emergency relief, the Federal Emergency Management Administration and the Small Business Administration, two of the federal agencies with the lowest staff morale at the time Katrina hit,[42] were restructured after Katrina and, working with state agencies, managed to improve some of their early responses to Sandy. The Army Corps of Engineers, Department of Transportation, Department of Housing and Urban Development (HUD), and other agencies also distributed billions of dollars that Congress appropriated for disaster relief and rebuilding after Sandy, even though many of these agencies took hits for responses perceived as slow, inadequate, or overly bureaucratic. The usual stories about delays in federal aid, state government incompetence, and contractor mismanagement appeared.

Some of the agencies promoting rebuilding incorporated measures to promote adaptation to coastal hazards, but in the United States and in the Bahamas, the focus on quick rebuilding extended existing policies that tolerate or even encourage people to build in the same hazardous sites, though perhaps with better structures. In the United States, agencies like HUD and policies like the mortgage tax deduction have for decades encouraged development and imposed few requirements that would have steered building away from low-lying or exposed sites along rivers and coasts. The Army Corps of Engineers had also built massive protective structures and had replenished beaches to protect coastal development.

The most important program affecting choices to live and invest in low-lying areas in the United States, the subsidized National Flood Insurance Program (NFIP), is discussed in chapter 11. Payouts from Hurricane Katrina had put this program into a massive deficit. Advocates who wanted to lower the deficit and reduce incentives for building in floodplains passed the Biggerts-Waters reforms into law in 2012, with exquisitely bad timing. Hurricane Sandy hit just before the first set of policyholders were to receive bills with higher premiums.[43] Insured owners facing damage from Hurricane Sandy formed protest groups in alliance with groups of the insured in other locations, and got most of the reforms rescinded or delayed in 2014. Policyholders also complained about slow payouts and denials of damage claims after Sandy. Investigations were opened to consider whether private insurers who administered the program had understated damage estimates related to Sandy in order to avoid FEMA audits that might require firms to reimburse the government for overpayments made to policyholders. FEMA reopened thousands of claims.[44]

As this brief attempt at insurance reform shows, changing policies that encourage development in hazardous areas cannot happen quickly. Even so, after Sandy, some national agencies in the United States began or extended programs to improve resilience and adaptation, some of which dated back to the 2010 BP oil spill in the Gulf, to Katrina, or even earlier.[45] The Army Corps of Engineers extended its slow adoption of methods other than levees and floodwalls with green infrastructure projects such as artificial wetlands and oyster reefs, which researchers find provide major reductions of damage from coastal storms.[46] The Department of Interior created grant programs to promote ecosystem-based protective functions in and near national protected lands in the Sandy region. The Sea Grant program of the National Oceanic and Atmospheric Administration sponsored grants to improve information about coastal storm hazards.[47] The Department of Housing and Urban Development sponsored a competition called Rebuild by Design, funding innovative landscape and engineering approaches to coastal areas in New Jersey and New York, a program that inspired similar plans and design competitions in San Francisco and other cities and a second, national round of competitive HUD grants.[48] Boston's experience of minor coastal flooding from Sandy was "more luck than planning." But seeing the damage in New York City, Boston designers, planners, and officials created an array of plans and designs, including novel ideas such as accommodating rising seas in ninety years by converting Back Bay streets into canals.[49]

Institutions with long-term agendas and often international reach linked Sandy to broad programs for climate adaptation, hazard reduction, or quality of life, often affirming that their planning so far has not been enough. The insurance industry has developed sophisticated modeling of climate change effects, created catastrophe bonds to provide immediate payouts (especially helpful for developing countries), and engaged in public outreach and policy discussions after Sandy. Professional societies such as the American Society of Civil Engineers, and national organizations in the United States such as the Association of State Floodplain Managers, renewed their ongoing calls for long-term investments in infrastructure and reducing development in floodplains. The Rockefeller Foundation created in 2013 an international funding initiative titled 100 Resilient Cities,[50] supplemented by a special set of projects focused on coastal cities in Asia. The most ambitious of these organizations' programs promote adaptation measures not simply as responses to climate change but also as means to recover quickly from routine storms, to strengthen democratic decision making, and to improve community life. Because hazards are caused jointly by structures, land forms, incentives, individual behavior, cultural values, and weather events, advocates see it as necessary to coordinate across institutions in government, nongovernmental organizations, and firms.[51]

For other institutions, the lack of planning for immediate emergencies, let alone long-term adaptation, became unfortunately clear in the United States and in the Caribbean. New Jersey Transit, one of the largest state-owned mass transit systems in the United States, stored dozens of engines and hundreds of cars in a low-lying rail yard in the waterfront city of Hoboken, and all of them were inundated by storm surge waters from the Hudson River.[52] The contractor managing all of the power lines for a New York state agency, the Long Island Power Authority, performed so poorly that Governor Cuomo signed legislation soon after Sandy privatizing its operations, while keeping the state board as a holding company.[53] The aid agencies operating camps for displaced persons in Haiti failed to prepare adequately for their flooding and for the subsequent increase in cholera cases.[54] As these organizations recovered, they tended to make changes quietly, rather than making public statements about the need to adapt, although New Jersey Transit's shift was highly public.[55]

THE CASES

In part 1, we discuss the difficulties of tracking and reporting this unusual storm and of ordering evacuations. The authors of these two chapters also describe the particular lessons that weather agencies and emergency managers took from Hurricane Sandy. In chapter 1, Steven Decker and New Jersey State Climatologist David Robinson explain how the combination of a large tropical storm with a cold front and blocking high pressure zone forced Hurricane Sandy into an unusual track, describing how forecasters updated their projections for its track and power as the storm progressed. Regarding whether Hurricane Sandy was transformational, Decker and Robinson report that the U.S. Weather Service has adjusted its forecasting models and changed its procedures to better communicate the message that a storm that is no longer labeled a hurricane can still be extremely destructive. Because climate change is loading the dice, increasing the likelihood of strong storms like Sandy, weather forecasters know that it will become increasingly important to monitor storm patterns and to send out effective warnings.

Daniel Baldwin Hess and Brian Conley (chapter 2) use Connecticut as a case study to explain how emergency managers weigh information such as transportation options, storm tracks, and residents' psychological responses to storms when they consider whether to order a locale to evacuate before a storm. Not only do officials have to decide on the basis of experience and forecasts whether the storm is likely to put people in their municipality in danger, but they also have to answer questions such as whether people in various neighborhoods can evacuate safely given current road and weather conditions and whether designating a large zone for mass departure might inadvertently jam evacuation routes and block the people who are truly in harm's way. Ideally, officials also make provisions for people who have no cars, the elderly, the infirm, and others who might not be able or willing to evacuate. Ordering an evacuation that proves to be unneeded is particularly worrisome, because people may be less likely to follow evacuation orders the next time. Sandy did substantial damage along Connecticut's coast, but with early storm tracking and recent experiences of storms and an evacuation drill, municipalities there reported that evacuations were successful. The procedures described in this chapter illustrate that emergency preparedness and execution tends to be well advanced in the United States; by contrast, chapters about land use illustrate that, as compared to other countries, hazard reduction is not well advanced in the United States.

Part 2 comprises case studies analyzing decisions in the days after the storm. Adelle Thomas (chapter 3) describes damage and reactions to the storm in the first places it hit—Cuba, Haiti, and the Bahamas—and she presents a detailed case study of two areas of the Bahamas that responded to Sandy very differently. Although people in the Bahamas tend to see their main island of New Providence as being largely safe from hurricanes, Hurricane Sandy damaged key roadways and other facilities serving the cruise ship harbor. Local political decisions to rebuild quickly, without changing the design of these facilities along the waterfront in New Providence, were a missed opportunity. Leaders on that island rushed to rebuild in order to reassure tourists that the beaches were ready for their next visit. People in the poorer and more rural outer islands, called the Family Islands, however, started a new round of plans to move people from hazardous areas and to adapt their infrastructure.

In chapter 4, Ashley Koning and David Redlawsk trace how the political fortunes of New Jersey governor Chris Christie, and, to a lesser extent, President Obama, were linked to attitudes about Hurricane Sandy, climate change, and recovery efforts. Their survey research through the Rutgers-Eagleton Institute poll tracks Christie's rise of popularity after the storm to a sustained 80 percent approval rate for his handling of Sandy, leading to his easy reelection and talk of a presidential run. Yet this was no simple narrative of a heroic political leader. A majority of people in New Jersey polled before Sandy were worried about climate change. Almost two-thirds polled after Sandy thought that climate change caused or contributed to the strength of Sandy. They were initially optimistic about government response efforts. Just over a year after the storm, though, Christie's favorability ratings declined dramatically, when he attempted to deflect accusations that his administration's staff closed lanes to a bridge into New York City, as retribution against political enemies, and again later, when evidence emerged that his administration used federal Sandy recovery aid to reward political allies. Even at the height of Christie's popularity, residents felt that it would take up to five years for the state to recover, and a large majority favored carefully assessing the likelihood of damage from future storms, rather than rebuilding immediately as promoted by Christie and other leaders.

Because extreme storms are a normal part of the environment, the word "disaster" is usually applied only when such storms harm humans. As Joanna Burger and Larry Niles explain in chapter 5, we can envision Sandy as a disaster for ecological systems as well, because much of Sandy's damage to the environment was made worse by human activities. Moving sand to build up beaches, building jetties and sea walls, and stringing roads along coastlines mean that the natural processes of shifting wetlands and coastal sand transport are disrupted. This chapter describes Sandy's damage to three types of ecosystems that are typical of the northeastern coastline in the United States. It then focuses on the sites of the hurricane's greatest ecological damage, beaches along the Delaware Bay that are critical to Red Knots, a threatened migratory shorebird species. Thousands of these birds usually arrive in the spring to feast on eggs laid by horseshoe crabs. An unusual combination of agencies, organizations, and volunteers came together before the following spring to replenish sand on these beaches, enabling the crabs to come onshore to lay their eggs. They are attempting to create permanent provision for ecological restoration as needed after future storms.

Chapter 6, by Angela Oberg, Julia Flagg, Patricia Clay, Lisa Colburn, and Bonnie McCay, discusses commercial fishermen who still work along the New York and New Jersey shore areas, even in areas now dominated by suburban development. As the largest and most diverse group of humans worldwide that relies on taking wild resources, fishermen are accustomed to uncertainty and are sensitive to environmental changes of all kinds. Fishermen can often adapt to declining stock by moving to new fishing sites or taking other types of fish. Interviews done soon after Sandy show that having to change fishing practices may reduce their income and make it difficult to pay for ongoing costs of keeping their boats and docks in order. One group of New Jersey fishermen that is the special focus of this chapter operates through a cooperative, and they expressed a greater sense of community resilience and interdependence than fishermen elsewhere in New Jersey and New York. The heavy damage their port in particular experienced may have increased their sense of group solidarity. Expressing the sentiment that they viewed fishing more as a way of life than as a business, these cooperative members saw the storm as one in a series of ongoing challenges that they have had to confront, and not as an event that was (negatively) transformational.

Case studies in part 3 examine how institutions are facing the future and whether they are planning for change. In chapter 7, Kenneth Gould and Tammy Lewis analyze the history of the Gowanus Canal in Brooklyn as it shifted from a creek hosting a rich diversity of species, to a closed and bulkhead-bound canal ringed by industries such as tanneries and coal gas plants, to a Superfund site, to the proposed location for luxury condominiums in an increasingly wealthy city under the administration of Mayor Michael Bloomberg. Flooding of local neighborhoods with toxic canal water during Hurricane Sandy had surprisingly little effect on plans to convert abandoned industrial lots around the canal into high-density condominiums. Gould and Lewis present a theory of green gentrification that predicts that low-income residents will be replaced by wealthier residents drawn to the "greened" environmental resource, despite the neighborhood's demonstrated vulnerability to climate change-induced flooding. They conclude that the political and economic forces driving the urban redevelopment growth machine in New York City is remarkably inured to flooding risks based on current conditions or those of climate change.

Mark Hewitt compares New Jersey shore resort towns' experience with Sandy to their experience of devastating fires and other disasters dating from the nineteenth century, in chapter 8, providing the long view of conditions on the coast and of the processes that drive rebuilding. These events have punctuated the history of tourist-led settlement along a coastline that is environmentally sensitive. Hewitt discusses contradictions among values for development, amusement, scenic beauty, and environmental protection that have shaped this region. Those who have been scared off from further investments on the shore after a disaster have generally been replaced by a new round of enthusiastic investors, sustaining or increasing the overall human exposure to hazards on the shore.

Explaining today's economic pressures to rebuild, in chapter 9 Briavel Holcomb describes the responses of people in the beach tourism industry in New Jersey and in Long Island, New York, who faced having to mount a tourism season within seven months of Hurricane Sandy. In both states, officials and industry leaders promoted the idea that the shore was open for business. Areas that depend on tourism often have few other economic activities and a short season for earning their year's income and so have little choice but to restart business as soon as possible. Relying on tourism has its costs, though. She explains that coastal areas that are dominated by seasonal residents and renters may have trouble mobilizing political support for actions promoting long-term resilience.

Clinton Andrews in chapter 10 considers how budgets in small municipalities, which dominate the Atlantic Coast shoreline in area if not in population, will fare as climate change progresses. Andrews uses data from three small New Jersey shore towns that differ demographically and geographically in an exercise of agent-based modeling, to project how property tax revenues would change after disasters. Andrews tests storms of varying sizes under three policy scenarios, all of which assume continued provision of the same level of municipal of services. One scenario, reflecting current policy, assumes that all properties will be rebuilt with substantial federal aid for cleanup and full coverage from the National Flood Insurance Program (NFIP). A second scenario assumes a change in federal policy so that only half of the owners receive federal insurance payouts or other subsidies to rebuild. A third scenario assumes that all of the properties with losses of more than 50 percent of their value are not rebuilt, either because of federal or state buyouts or abandonment by owners, a policy option that has not been widely supported by states or national politicians. The third scenario results in the smallest increase in local property tax rates, a result that would likely surprise most local officials. The second scenario where the federal government reduces its subsidies for flood insurance and cleanup provokes much higher local property tax rates. All three policy options show likely increases in property tax rates and call into question how municipalities in the most vulnerable locations will sustain their services as the climate changes.

Officials aiming to reduce people's exposure to hazards must work quickly after a disaster because individuals and municipal governments make decisions immediately following a disaster that guide eventual rebuilding plans. Mariana Leckner, Melanie McDermott, James Mitchell, and Karen O'Neill report in chapter 11 how three small municipalities in New Jersey are making decisions as they recover from Hurricane Sandy. Focus group participants and local officials reported that federal policies assuming that rebuilding decisions happen long after the initial recovery and cleanup stages did not match up with local needs. One town, with moderate average household income and damage to 10 percent of its taxable property, has depended heavily on outside help from agencies and private charities, and its future landscape and policy choices remain unclear. The other two towns, with higher average incomes and damage ranging from 3 to 5 percent of their taxable properties, have been better able to navigate federal and state aid policies and have begun to talk about possible changes in land use policies, although they have not made major changes to reduce their exposure to hazards.

Whether a community located in a hazardous area remains viable may come to depend in part on the ability of utilities to continue providing services at affordable costs. In chapter 12, Daniel Van Abs describes the systems for water, wastewater, and stormwater that are typical in the coastal United States. He outlines their vulnerabilities to storms and sea level rise, illustrating these vulnerabilities with a survey of water utility professionals in New Jersey and a review of damage after Sandy in New Jersey and in New York City. Sewage treatment plants, pipes, and pumps, many of which are necessarily sited along the coast, were especially vulnerable. He then describes how one water utility's pilot project that used a global positioning system (GPS) to map water service lines allowed it to quickly stop leaks and to avoid billing users whose houses were damaged. The loss of customers affects the feasibility and cost of offering utility services. Improved tracking of utility use may change the calculations of utility providers, as they look at the costs of rebuilding after disasters. Regulated water utilities are required to provide services to eligible development, even in highly hazardous areas like barrier islands, and so service will continue until other drivers change the pattern and existence of development in at-risk areas, with ratepayers paying the costs.

In chapter 13, Frank Felder and Shankar Chandramowli discuss the challenges of an aging electric grid and disincentives and incentives for improvements faced by regulators and utilities in New Jersey and New York State. They discuss how electric service has been the foundation of economic development, explain how power generation and distribution systems are vulnerable to climate change, and describe the disruptions that Sandy caused to all sectors and the reactions of institutions. They then outline changes the sector could make to improve overall system performance today and to reduce and respond to climate change as it progresses. Regulators are reassessing the effects of climate change, in part in reaction to Hurricane Sandy, but regional utilities have just begun to make new investments and policy changes that Felder and Chandramowli see as necessary in the face of climate change.[56]

The conclusion returns to our original questions and addresses overall findings about the changes that have been made and the problems that remain and lessons from these case studies. It examines how the case studies answered our central question, of whether Hurricane Sandy was transformational, by looking at the outcomes of several supporting questions that we had to deal with first. Have organizations and individuals changed their aims and practices in response to Sandy because they have changed their expectations about risk and perceptions of coastal vulnerability? What plans or actions do they have under way? What other forcing factors or constraints do people face? What are the cultural or emotional elements of their responses? And what do they say about how they and others should react to Sandy in the long run? The conclusion also notes potential responses to climate hazards that have received little attention since the storm.

Human responses to severe coastal storms and to climate change in the near future will almost certainly involve a messy combination of inaction, making do, and careful efforts to reduce specific hazards along the urbanized coast. Some organizations use science as they make decisions, but the more immediate sources of information for most people are personal finances and preferences, budget projections, political calculations, and anecdotes about previous storms. Residents in every country with a coastline will pay, regardless of the choices that are made.

NOTES

1. National Weather Service, "Hurricane Sandy—October 29, 2012," accessed October 4, 2014, http://www.weather.gov/okx/HurricaneSandy.
2. Robert J. Brulle, "Institutionalizing Delay: Foundation Funding and the Creation of U.S. Climate Change Counter-movement Organizations," *Climatic Change* 122, no. 4 (2014): 681–694.
3. Andrew Kohut et al., "More Say There Is Solid Evidence of Global Warming" (Washington, DC: Pew Research Center for the People and the Press, 2012), http://www.people-press.org/files/legacy-pdf/10-15-12%20Global%20Warming%20Release.pdf; Anthony Leiserowitz, "Climate Change Risk Perception and Policy Preferences: The Role of Affect, Imagery, and Values," *Climatic Change* 77, nos. 1–2 (2006): 45–72.
4. The cover, which appeared on newsstand copies, was widely reproduced, including Eyder Peralta, "*Bloomberg Businessweek*'s Cover: 'It's Global Warming, Stupid,'" The Two-Way: Breaking News From NPR, http://www.npr.org/sections/thetwo-way/2012/11/01/164106889/bloomberg-businessweeks-cover-its-global-warming-stupid. The issue's cover article is Paul M. Barrett, "It's Global Warming, Stupid," *Bloomberg Businessweek*, November 1, 2012, http://www.businessweek.com/articles/2012-11-01/its-global-warming-stupid.
5. Jimmy Vielkind, "Cuomo: 'Climate Change Is a Reality . . . We Are Vulnerable,'" *Albany Times-Union*, October 31, 2012, http://blog.timesunion.com/capitol/archives/162798/cuomo-climate-change-is-a-reality-we-are-vulnerable.
6. Alice Newton and Juergen Weichselgartner, "Hotspots of Coastal Vulnerability: A DPSIR Analysis to Find Societal Pathways and Responses," *Estuarine, Coastal and Shelf Science* 140 (2014): 123–133.

7. Robert W. Kates, William R. Travis, and Thomas J. Wilbanks, "Transformational Adaptation When Incremental Adaptations to Climate Change Are Insufficient," *Proceedings of the National Academy of Sciences* 109, no. 19 (2012): 7156–7161.
8. Herman Gerritsen, "What Happened in 1953? The Big Flood in the Netherlands in Retrospect," *Philosophical Transactions of the Royal Society—A* 363 (2005): 1271–1291.
9. Delaware River Basin Commission, *2011 Annual Report: Celebrating Our 50th Anniversary* (West Trenton, NJ: DRBC, 2011), http://www.state.nj.us/drbc/library/documents/2011AR.pdf; Becky Oskin, "Hellish Northridge Earthquake: Is Los Angeles Safer 20 Years Later?," *Live Science*, January 17, 2014, http://www.livescience.com/42675-northridge-earthquake-20th-anniversary-science.html; Adrian Sainz, "Florida Building Codes, Revamped since Andrew, Still Being Worked," *Insurance Journal*, May 18, 2007, http://www.insurancejournal.com/news/southeast/2007/05/18/79827.htm.
10. Federal Emergency Management Administration, *National Flood Insurance Program: Program Description* (Washington, DC: FEMA): 1–4, http://www.fema.gov/media-library-data/20130726-1447-20490-2156/nfipdescrip_1_.pdf; Erwann O. Michel-Kerjan, "Catastrophe Economics: The National Flood Insurance Program," *Journal of Economic Perspectives* 24, no. 4 (2010): 27–35.
11. U.S. Department of Homeland Security, "Creation of the Department of Homeland Security," accessed May 15, 2015, http://www.dhs.gov/creation-department-homeland-security.
12. John McCormick, "Now on Higher Ground, Ex-river Town Thriving: Some Say '93 Flood Proved to Be Good for Valmeyer, Ill.," *Chicago Tribune*, July, 7, 2003, http://articles.chicagotribune.com/2003-07-07/news/0307070177_1_higher-ground-valmeyer-great-flood.
13. American Society of Civil Engineers, *The New Orleans Hurricane Protection System: What Went Wrong and Why: A Report by the American Society of Civil Engineers Hurricane Katrina External Review Panel* (Reston, VA: ASCE, 2007), http://www.asce.org/uploadedfiles/publications/asce_news/2009/04_april/erpreport.pdf.
14. Don Jergler, "New FEMA Guidelines Force the Climate Change Issue," *Insurance Journal*, March 26, 2015, http://www.insurancejournal.com/news/national/2015/03/26/362248.htm.
15. National Research Council, *Disaster Resilience: A National Imperative* (Washington, DC: National Academies Press, 2012), http://www.nap.edu/catalog.php?record_id=13457; James C. Schwab, ed., *Hazard Mitigation: Integrating Best Practices into Planning* (American Planning Association, 2010), http://www.fema.gov/media-library/assets/documents/19261.
16. Stephane Hallegatte et al., "Future Flood Losses in Major Coastal Cities," *Nature Climate Change* 3, no. 9 (2013): 802–806; Robert E. Kopp et al., "Probabilistic 21st and 22nd Century Sea-Level Projections at a Global Network of Tide-Gauge Sites," *Earth's Future* 2, no. 8 (2014): 383–406.
17. John R. Gillis, *The Human Shore: Seacoasts in History* (Chicago: University of Chicago Press, 2012).

18. Julian Agyeman, Patrick Devine-Wright, and Julia Prange, "Close to the Edge, Down by the River? Joining Up Managed Retreat and Place Attachment in a Climate Changed World," *Environment and Planning A* 41, no. 3 (2009): 509–513.
19. Gordon McGranahan, Deborah Balk, and Bridget Anderson, "The Rising Tide: Assessing the Risks of Climate Change and Human Settlements in Low Elevation Coastal Zones," *Environment and Urbanization* 19, no. 1 (2007): 17–37.
20. Hallegatte et al., "Future Flood Losses in Major Coastal Cities."
21. Rona Kobell, "For Vulnerable Barrier Islands, a Rush to Rebuild on U.S. Coast," *Yale Environment* 360 (2015): accessed March 1, 2015, http://e360.yale.edu/feature/for_vulnerable_barrier_islands_a_rush_to_rebuild_on_us_coast/2838; Christopher Turbott, "Managed Retreat from Coastal Hazards: Options for Implementation" (Hamilton East, NZ: Andrew Steward Ltd., for Environment Waikato Regional Council, New Zealand, 2006).
22. McGranahan, Balk, and Anderson, "The Rising Tide."
23. Peter Stalker, ed., *Technologies for Adaptation to Climate Change* (Bonn: United Nations Climate Change Secretariat, 2006).
24. Edward Lynch, "Hurricane Sandy's Wake-Up Call: The New York Area Redefines Recovery," *Planning* (February 2015).
25. Luciana S. Esteves, "Is Managed Realignment a Sustainable Long-Term Coastal Management Approach?," in special issue, *Journal of Coastal Research* 65 (2013): 933–938.
26. Lykke Andersen, Lotte Lund, and Dorte Verner, "Migration and Climate Change," in *Reducing Poverty, Protecting Livelihoods, and Building Assets in a Changing Climate: Social Implications of Climate Change for Latin America and the Caribbean*, ed. Dorte Verner (Washington, DC: World Bank, 2010), 195–220.
27. Alliance of Small Island States, accessed March 1, 2015, http://aosis.org; Conner Bailey, Robert Gramling, and Shirley B. Laska, "Complexities of Resilience: Adaptation and Change within Human Communities of Coastal Louisiana," in *Perspectives on the Restoration of the Mississippi Delta*, ed. John W. Day et al. (Dordrecht: Springer Netherlands, 2014), 125–140.
28. Andersen, Lund, and Verner, "Migration and Climate Change."
29. We are indebted to an anonymous peer-reviewer for this latter insight.
30. For example, in the United Kingdom, flooding in early 2014 raised significant concerns about the capacity of government agencies to act; see "High Water Everywhere," *Economist*, February 15, 2014, http://www.economist.com/news/britain/21596535-ordinary-britons-have-so-far-coped-admirably-widespread-flooding-rain-still.
31. Mark Waller, "Hurricane Katrina Eight Years Later, a Statistical Snapshot of the New Orleans Area," *Times-Picayune*, August 28, 2013, http://www.nola.com/katrina/index.ssf/2013/08/hurricane_katrina_eight_years.html.
32. For social scientific analyses of Katrina, see David L. Brunsma, David Overfelt, and J. Steven Picou, eds., *The Sociology of Katrina: Perspectives on a Modern Catastrophe* (Lanham, MD: Rowman and Littlefield, 2007); William R. Freudenburg et al., *Catastrophe in the Making* (Washington, DC: Island Press/Shearwater Books, 2009).
33. David S. Heller, "Evacuation Planning in the Aftermath of Katrina: Lessons Learned," *Risk, Hazards and Crisis in Public Policy* 1, no. 2 (2010): 131–174.

34. Katy Reckdahl, "Why New Orleans' Katrina Evacuation Debacle Will Never Happen Again," Next City, July 9, 2014, http://nextcity.org/daily/entry/new-orleans-evacuation-hurricane-katrina-will-never-happen-again.

35. Tulip Mazumdar, "Typhoon Haiyan Death Toll Rises over 5,000," BBC.com, November 22, 2013, http://www.bbc.com/news/world-asia-25051606.

36. New York City, *PlaNYC: A Greener, Greater New York* (New York City, 2007, updated 2011), http://www.nyc.gov/html/planyc2030/html/home/home.shtml.

37. New York City, *A Stronger, More Resilient New York* (New York City, 2013), http://www.nyc.gov/html/planyc2030/html/home/home.shtml.

38. Sandy Regional Assembly, "SIRR Analysis," (2013), http://goodjobsny.org/sites/default/files/docs/sandy_regional_assembly_sirrreport_hstsrecommendations_072213_final.pdf.

39. New York City, "De Blasio Administration Releases PlaNYC Progress Report, Highlighting Major Accomplishments on Sustainability and Resiliency Efforts," April 22, 2014, http://www1.nyc.gov/office-of-the-mayor/news/167–14/de-blasio-administration-releases-planyc-progress-report-highlighting-major-accomplishments-on.

40. Andrew M. Cuomo, "NY Rising: 2013 State of the State," State of New York, New York Rising: Community Reconstruction Program (2013), http://stormrecovery.ny.gov/community-reconstruction-program.

41. State of New Jersey, "Governor's Office of Rebuilding and Recovery," accessed March 1, 2015, http://nj.gov/gorr.

42. FEMA was surveyed as part of the Department of Homeland Security, "Federal Human Capital Survey" (Washington, DC: U.S. Office of Personnel Management, 2004), http://www.fedview.opm.gov/2004.

43. Federal Emergency Management Agency, "Biggert-Waters Flood Insurance Reform Act of 2012 (BW12) Timeline," accessed May 10, 2015, http://www.fema.gov/media-library-data/20130726–1912–25045–8239/bw_timeline_table_04172013.pdf; Federal Emergency Management Agency, "Flood Insurance Reform: The Law," accessed May 10, 2015, http://www.fema.gov/flood-insurance-reform-law.

44. Coral Davenport, "Popular Flood Insurance Law Is Target of Both Political Parties," *New York Times*, January 29, 2014; Charles Lane, "Why Sandy Homeowners Were Left in the Lurch," WNYC, May 14, 2015, http://www.wnyc.org/story/why-sandy-homeowners-were-left-lurch; Michel-Kerjan, "Catastrophe Economics: The National Flood Insurance Program"; Christie Smythe, "Sandy Insurance Issues Said to Be Focus of Criminal Probe," *Bloomberg Business*, January 24, 2015, http://www.bloomberg.com/news/articles/2015–01–24/sandy-insurance-issues-said-to-be-focus-of-n-y-criminal-probe; Michel-Kerjan notes that the program was designed from the start to cover most disasters but that for the largest disasters, the program would always need additional funding from the federal government.

45. Hurricane Sandy Rebuilding Task Force, *The Hurricane Sandy Rebuilding Strategy* (Washington, DC: U.S. Department of Housing and Urban Development, HSRT, 2013), 49, http://portal.hud.gov/hudportal/documents/huddoc?id=hsrebuildingstrategy.pdf.

46. Katie K. Arkema et al., "Coastal Habitats Shield People and Property from Sea-Level Rise and Storms," *Nature Climate Change* 3 (2013): 913–918, http://www.nature.com/nclimate/journal/v3/n10/full/nclimate1944.html.

47. "Coastal Storm Awareness Program," 2014 Sea Grant, National Oceanic and Atmospheric Administration, http://seagrant.noaa.gov/FundingFellowships/CoastalStormsAwarenessProgram.aspx; Hurricane Sandy Coastal Resiliency Competitive Grant Program, administered by the National Fish and Wildlife Foundation, accessed March 10, 2013, http://www.nfwf.org/hurricanesandy/Pages/home.aspx#.U5zGcfldWPs; U.S. Department of Housing and Urban Development, "Rebuild by Design," accessed May 31, 2014, http://portal.hud.gov/hudportal/HUD?src=/sandyrebuilding/rebuildbydesign.

48. Julie Wormser, "A Tale of Two Cities: Boston, New York, and Hurricane Sandy," Boston Harbor Association, November 11, 2012, http://tbha.org/blog/jwormser/tale-two-cities-boston-new-york-and-hurricane-sandy.

49. Virginia Quinn, "The Urban Implications of Living with Water" (Boston: Urban Land Institute Boston / New England District Council, Kresge Foundation, 2014), http://boston.uli.org/wp-content/uploads/sites/12/2012/04/ULI_LivingWithWater-Final1.pdf; Kevin Wilcox, "Guide Aims to Bolster Boston's Resilience," *Civil Engineering*, November 19, 2013, http://www.asce.org/cemagazine/Article.aspx?id=23622328676#.VFS7KvnF8nU.

50. "Perilous Paper: Bonds That Pay Out When Catastrophe Strikes Are Rising in Popularity," *Economist*, October 5, 2013, http://www.economist.com/news/finance-and-economics/21587229-bonds-pay-out-when-catastrophe-strikes-are-rising-popularity-perilous-paper; Rockefeller Foundation, "100 Resilient Cities: About the Challenge," accessed May 31, 2014, http://100resilientcities.rockefellerfoundation.org/pages/about-the-challenge.

51. Lynch, "Hurricane Sandy's Wake-Up Call."

52. Andrea Bernstein and Kate Hinds, "How New Jersey Transit Failed Sandy's Test," WNYC/*Record*, May 13, 2013, http://www.wnyc.org/story/292666-njtransit-sandy/#.

53. State of New York, "Governor Cuomo Signs Legislation Restructuring Utility Operations on Long Island," September 29, 2013, http://www.governor.ny.gov/press/07292013-restructure-utility-operations-on-long-island.

54. United Nations Office for the Coordination of Humanitarian Affairs, "Haiti: Hundreds of Thousands of People Affected by Hurricane Sandy," November 2, 2012, http://www.unocha.org/top-stories/all-stories/haiti-hundreds-thousands-people-affected-hurricane-sandy.

55. Karen Rouse, "N.J. Transit Unveils New Plan to Protect Equipment in Severe Weather," *Bergen Record*, June 11 2014, http://www.northjersey.com/news/nj-transit-unveils-new-plan-to-protect-equipment-in-severe-weather-1.1033311.

56. Alexi Friedman, "PSE&G Reaches Agreement on $1.2 Billion 'Energy Strong' Grid-Hardening Project," *Star Ledger*, May 1 2014, http://www.nj.com/business/index.ssf/2014/05/pseg_reaches_agreement_for_scaled-down_version_of_massive_grid-hardening_project.html.

PART I THE STORM

1 • HURRICANE SANDY FROM METEOROLOGICAL AND CLIMATOLOGICAL PERSPECTIVES

STEVEN G. DECKER AND
DAVID A. ROBINSON

Hurricane Sandy and its post-tropical remnants dealt portions of the Caribbean and a broad swath of the eastern United States a punishing blow. This late October storm brought powerful winds, a massive storm surge, flooding rains and even smothering snow that damaged or destroyed thousands of buildings, brought extended power outages to many millions of customers, disrupted transportation, and tragically, killed several hundred people.[1] Damage from Sandy in the United States totaled approximately $50 billion, making this the second-most costly storm on record, after Hurricane Katrina in 2005.[2] While weather records and modeling indicate that Sandy was not the worst storm that could have hit areas along its track, its exceptional size and unusual path, and its target of one of the most populous portions of the United States, made it an exceptionally memorable event. One can consider it at least as a physically transformative event for the worst-hit areas. It also quickly led to a change in the way storm warnings are made at the U.S. National Hurricane Center (NHC), improving information that is used across the Caribbean and North America.

As an unusual storm affecting places outside the usual hazard zones, Sandy illustrates why the international community of forecasters routinely reviews and changes the way they forecast and communicate about storms. This chapter examines the life history of the storm from its formation in the Caribbean to its unusual landfall on the New Jersey coast. We explain its path, size, and strength; assess the success in forecasting its path; and explain changes in warning practices at the National Hurricane Center.

THE EVOLUTION OF SANDY FROM A METEOROLOGICAL PERSPECTIVE

The tropical depression that would become Hurricane Sandy formed over the warm waters of the western Caribbean Sea at 2 P.M. Eastern Daylight Time (EDT) on October 21, 2012, and strengthened to hurricane status by 8 A.M. EDT October 24, 2012, ninety miles south of Jamaica. Hurricane Sandy tracked north, making landfall and causing damage in Jamaica and Cuba along the way.[3] Its track between these landfalls is included in figure 1.1. Sandy briefly weakened below hurricane status in its first westward jog near the Bahamas between 8 P.M. October 26 and 8 A.M. October 27, strengthened during its second westward jog, and slowly weakened as it lost its tropical characteristics just before making landfall near Brigantine, New Jersey, with estimated sustained winds of 81 miles per hour offshore and an estimated central pressure of 27.91 inches of mercury according to the National Hurricane Center's best track interpolation.[4] (Observed winds and pressures around the time of landfall were not quite as extreme.)

Complex circumstances made the storm unusually wide and its path toward the East Coast of the United States difficult to predict at first, although accurate predictions soon emerged. We next describe some of those circumstances, including Sandy's formation in the Caribbean and its interactions with midlatitude circulations.

Tropical Stage

The western Caribbean is one of the most likely places to see tropical cyclone formation in October, so in this regard Sandy was not unusual. Hurricanes forming there typically track either west into Central America, northwest into the Gulf of Mexico, or north, threatening Florida before turning northeast and back out to sea. Sandy followed the third track, at least at first, until a series of interactions with extratropical weather systems led to a more atypical evolution.

Midlatitude Interaction

Sandy's track was unusual and difficult to predict because patterns in midlatitude winds were unusual. The general circulation in the midlatitudes consists of westerly winds, and so most tropical cyclones from the Caribbean eventually turn northeastward out to sea. These westerlies do not blow purely from west to east, but instead undulate in a wavelike pattern. The tops (that is, those closer to the poles) of the waves are called ridges, and the bottoms (that is, closer to the equator) are called troughs, and these are also characterized by lower and higher pressures, respectively, than average for the region. The wavelike pattern of wind flow during Sandy is shown in figure 1.1.[5] The top panel image indicates the large-scale flow when Sandy was near Jamaica, and the bottom panel image shows the flow around the time Sandy made landfall in New Jersey. In the top panel, the medium-gray band that starts at the California-Oregon border and undulates to the south around Utah is a trough, indicated by the letter A. The northward undulation to Wisconsin is a ridge, followed by another southward undulation marking the next trough, indicated by the B in Alabama.

The configuration of the midlatitude flows steered Sandy throughout its lifetime.[6] The atmospheric circulation at 8 P.M. EDT on October 24, 2012, summarized in the top panel of figure 1.1, contains a number of features that would influence Sandy's track over the next five days. Upper-level troughs A, B, and C are associated with cyclonic (counterclockwise in the Northern Hemisphere) circulations aloft. Upper-level ridge D is associated with an anticyclonic (clockwise) circulation aloft. Some early computer-generated weather forecasts showed that Sandy would be directed out to sea by the northwesterly winds found west and south of trough C. As the storm advanced, though, troughs A and B along with ridge D would eventually steer Sandy toward New Jersey.

Sandy first interacted with trough B. As Sandy moved north of Cuba, the southerly and southeasterly winds circulating around B started to draw Sandy northwest toward the Bahamas. This weakened Sandy below hurricane status, as strong winds aloft over the storm temporarily increased, attempting to rip Sandy apart. However, rather than falling apart, Sandy was

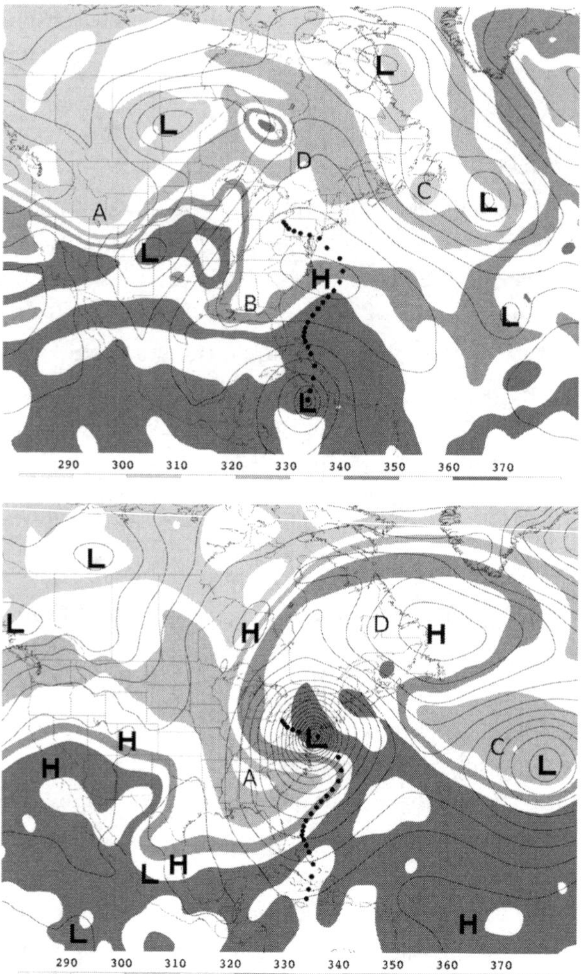

FIGURE 1.1. The tropospheric circulation at 8:00 p.m., October 24, 2012 (*top*), and 8:00 p.m., October 29 (*bottom*), with dotted lines indicating Sandy's position at six-hour increments. Circulations near the level of the jet stream are characterized by potential temperature at the tropopause (in degrees K, shaded according to legend). Higher temperatures (darker shades) represent upper-level ridges and upper-level anticyclones as well as tropical air masses. Lower temperatures (lighter shades) represent upper-level troughs and upper-level cyclones as well as polar air masses. Circulations in the lower troposphere are characterized by 850-hPa geopotential height contours (contoured every ninety-eight feet; relative highs/lows marked by H/L, respectively). Lower heights represent cyclones, whereas higher heights represent anticyclones. (Source: Data are derived from Global Forecast System (GFS) model analyses.)

able to ingest trough B into its own circulation, strengthening once again as it resumed a more north-to-northeast track, and, perhaps most important, doubled in size to more than 230 miles across.

Meanwhile, troughs A and C and ridge D had slowly propagated eastward and strengthened. The counterclockwise circulation around A combined with the clockwise circulation around D helped push Sandy toward the northwest and its eventual encounter with the New Jersey coast.

Post-Tropical Transition

As Sandy began to interact with trough A, it passed over the Gulf Stream, which allowed it to briefly strengthen to Category 2 status (sustained wind between 96 and 110 mph) less than a day before landfall. A complex interplay between the trough, Sandy, and the Gulf Stream then led to the rapid loss of Sandy's tropical characteristics. It was declared post-tropical, because it no longer had the defining features of a hurricane, at 5 P.M. EDT on October 29, 2012, two and a half hours before making final landfall. However, in this case the loss of tropical characteristics did not involve a rapid weakening of the storm. For one thing, following twenty-one consecutive months of above-normal temperatures in the region, sea surface temperatures were above average. If the sea surface temperatures had been normal, the storm would have taken the same track but would have been weaker as it headed toward a landfall.[7] Also, Sandy grew even larger during this second trough interaction. By the time Sandy was over Pennsylvania, it had been absorbed into the center of the circulation associated with trough A. The bottom panel image in figure 1.1 shows the result of this transition soon after landfall, as Sandy was steered northwestward by trough A and ridge D.

Although trough C did not direct Sandy out to sea, the lower-tropospheric cyclone (that is, the circular region of low heights indicated by the "L" east of trough C) associated with that trough provided for a long distance, east-to-west fetch of strong winds across the Atlantic Ocean. These winds were directed at the coasts of New Jersey, New York, and New England, when combined with the circulations around Sandy and the anticyclone to its northeast. This process, in conjunction with Sandy's large size, helped produce an extreme storm surge.[8]

To summarize, Sandy's unusual track was the result of a series of complex interactions with troughs and ridges in the midlatitude flow, culminating in Sandy's movement northwestward around a counterclockwise-rotating trough centered in Alabama and a clockwise-rotating ridge over Labrador.

Forecasting Sandy

Despite its unusual track and complex post-hurricane transition, numerous weather models noted at a remarkably early stage the potential for Sandy to menace the Caribbean and the East Coast. From the moment Sandy formed, it was accurately forecast to make landfall in Jamaica and Cuba two to three days later, although its intensity was initially under-forecast. Any landfall along the East Coast was more than a week away at that point.

Particularly at long lead times, a forecasting approach using numerous model runs (an ensemble) can provide an informative range of possible weather outcomes. In this section, we examine the output of the Global Ensemble Forecast System (GEFS) run by the U.S. National Weather Service (NWS).[9] In ensemble forecasting, multiple numerical predictions are made using slightly different initial conditions. With GEFS, every six hours (or cycle), a series of twenty two-week forecasts is generated. Each of the twenty ensemble members starts from slightly different but equally plausible initial atmospheric conditions to account for uncertainty in the state of the atmosphere at the initial time. Because the atmosphere is a chaotic system, even small changes in the atmospheric circulation can lead to completely different forecasts days later. By examining the ensemble members, one can examine a variety of possible outcomes and attach probabilities to them.

Each cycle of the GEFS was examined starting eleven days prior to landfall in New Jersey (the 8 P.M. EDT October 18 cycle) to see how many ensemble members showed Sandy making landfall somewhere in the United States or in New Jersey in particular (figure 1.2). The first indication a landfall was possible occurred during the 2 P.M. EST October 20, 2012, cycle (ten days prior to landfall and two days prior to a tropical depression forming in the Caribbean), when 10 percent of the ensemble members indicated such an occurrence. Remarkably, one ensemble member forecast a New Jersey landfall eight and a half days out (six hours prior to depression formation). Over the next few days, a U.S. landfall remained a fairly low probability event. Most ensemble members showed trough C directing Sandy out to sea. However, roughly six to four days prior to landfall, the probability of a U.S. landfall increased substantially. Between four and two

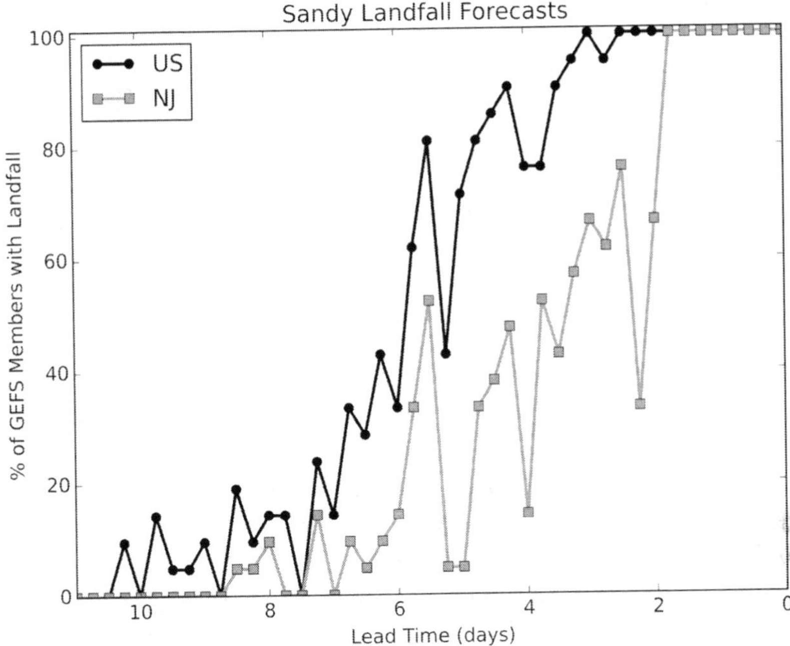

FIGURE 1.2. The percentage of projected Hurricane Sandy tracks from the GEFS that showed landfall in the United States (circles) or New Jersey (squares) as a function of lead time.

days prior to landfall, a New Jersey landfall became more and more likely. Finally, by Saturday night (October 27, leading into October 28), a New Jersey landfall became a virtual certainty, as every ensemble member in a series of GEFS cycles showed such a track. Less certain were the strength of Sandy and the timing by which it would lose tropical characteristics. This led to some confusion as to how the storm would be labeled and communicated in public forecasts made by the NWS, which we discuss later in the context of Sandy's transformative impact. (See chapter 2 by Daniel Hess and Brian Conley concerning how emergency managers used these early forecasts to make evacuation orders.)

Although not analyzed in depth here, a similar ensemble forecasting system developed by the Europeans also provided sets of forecasts of Sandy's track. Those forecasts gained attention when they accurately (as it turned out) showed relatively high probabilities of a U.S. landfall seven or eight days out, several days before the GEFS.[10] Less widely publicized was that the United Kingdom's model performed just as well if not better than the European model. On the other hand, the U.S.-based GEFS forecasts were more accurate when the time to landfall was less than three days. Nevertheless, the relatively poor performance of the GEFS a week before landfall helped the U.S. NWS to justify obtaining funds for a new supercomputer to facilitate better modeling.[11]

SANDY STRIKES NEW JERSEY

Atmospherically, Sandy's effects were felt over most of the eastern United States, with high wind warnings issued as far west as Lake Michigan, several feet of wet snow accumulating in the mountains of West Virginia, and severe thunderstorm warnings posted in eastern Massachusetts. On the morning of October 30, clouds associated with the post-tropical cyclone extended from South Carolina to southern Greenland. Sandy's storm surge severely inundated and damaged areas from Delaware Bay northward into southern New England. New Jersey sat at the epicenter of Sandy's wind, rain, and surge. Maps in figure 1.3 depict extremes in atmospheric pressure, wind gusts, and precipitation observed within the state during the storm. Most pressure and wind observations are gleaned from the Office of the New Jersey State Climatologist's (ONJSC) fifty-two-station New Jersey Weather and Climate Network (NJWxNet). Observations from these stations were received at the ONJSC every five minutes throughout Sandy. Mainly powered by solar-charged batteries and transmitting observations via cellular modem, all but six stations (which were run on AC power systems that failed) reported at the height of the storm. Only one anemometer was storm-damaged in New Jersey.

Barometric Pressure

The Atlantic City National Ocean Service (NOS) station reported an atmospheric pressure reading of 27.92 inches at time of landfall. This is 0.02 inches lower than the minimum pressure recorded on land during the deadly, devastating 1938 hurricane that hit Long Island and southeast New England, and it is the lowest pressure ever observed north of Cape Hatteras, North Carolina. The National Weather Service Atlantic City Airport station several miles inland in Pomona reported 28.00 inches. Both Atlantic

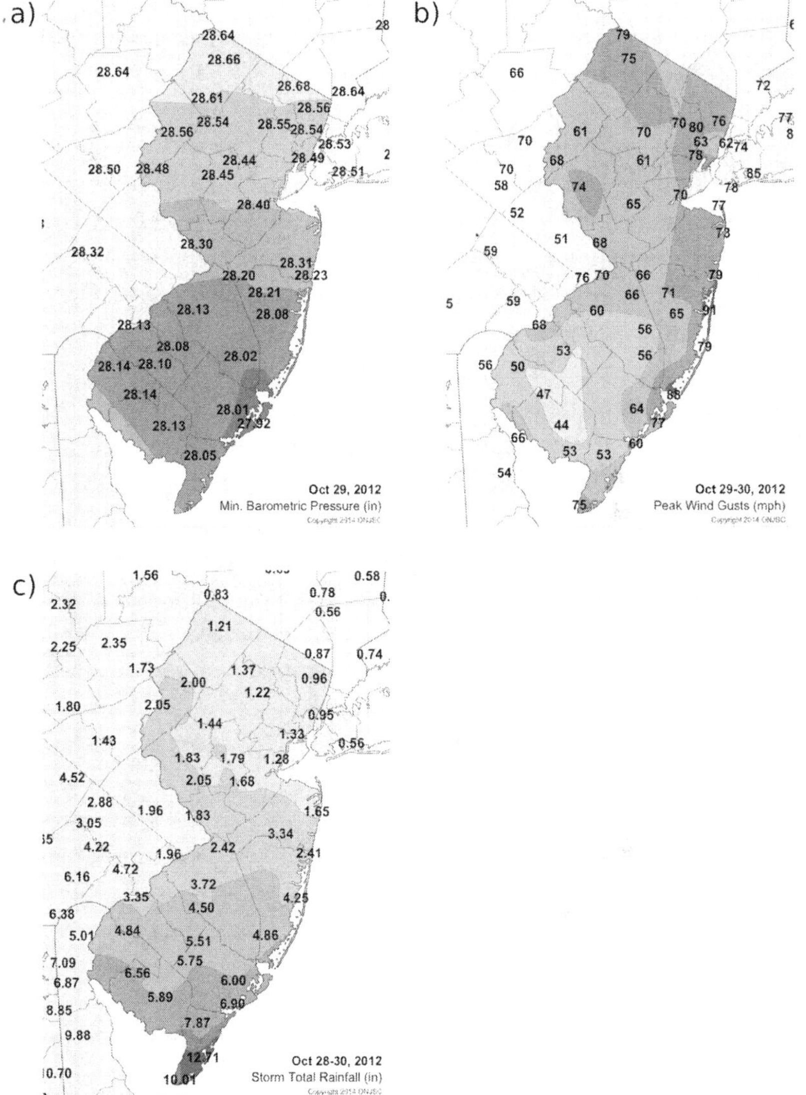

FIGURE 1.3. Sandy extremes: a) minimum pressures, b) maximum wind gusts, and c) total precipitation. (Sources: Pressure and wind observations are from NJ Weather and Climate Network observations (NJWxNet). Precipitation is from NJWxNet; NJ Community Collaborative Rain, Hail, and Snow network; and NWS Cooperative Observing Program stations.)

City area observations are well below the previous land-based state record of 28.36 inches in Long Branch during a nor'easter on March 6, 1932. During the evening of October 29, all of southern New Jersey exceeded the previous single-station records.

Wind

One of the more vexing issues in the post-Sandy analysis of winds recorded at weather stations is determining the speeds that sustained winds (averaged over at least two minutes) and gusts (generally two-second observations) reached during the storm. Anemometers exposed in open fields or at highly elevated sites will record winds of greater intensity than those in lower and more sheltered locations. Anemometers placed on a rooftop or between buildings may also record faster winds because the structures can create funneling effects. Despite the differences in anemometer type, recording practice, and siting, it is fortunate that observations were available from many stations. This helped facilitate quality control, determining which station observations were best representative of wind for a given type of exposure (surrounding and elevation) and rendering suspicious any observation that greatly exceeded others in a local area.

It can be stated with some confidence that maximum sustained winds were well over tropical storm force (39 mph) at coastal locations and at exposed inland locations in northern and central New Jersey, though they did not appear to exceed hurricane force (74 mph) at any location. Gusts exceeded hurricane force at many coastal locations and at some exposed inland sites. Within the NJWxNet, Seaside Heights, on a barrier island, had a peak gust of 91 mph. This was followed by 79 mph gusts at both Sea Girt, on the coast, and at High Point Monument, far inland at the northwest corner of the state. Other wind observations that were measured with quality instruments in well-exposed coastal locations include gusts as high as 88 mph at Tuckerton and 87 mph at Sandy Hook. Only coastal locations in the south, along both the Atlantic Ocean and Delaware Bay, received gusts approaching or just exceeding hurricane force.

Precipitation

Cape May, the southernmost county in New Jersey, took top honors in terms of Sandy precipitation. From the morning of October 28 through the early afternoon of the 30th, Stone Harbor led the way with 12.71 inches, followed by 11.91 inches and 11.70 inches at two Wildwood Crest locations. Totals fell off to the north, with some Bergen County locations in northeastern New Jersey receiving less than 1.00 inch. While precipitation was not excessive over most of the state, the far southern deluge statistically has a 0.5 percent per year probability of occurrence, otherwise described as a two hundred–year storm. Thus Sandy followed the general climatological rule that the left side (relative to the path) of a tropical storm, once into the middle latitudes, produces the most rainfall, while winds are strongest on the right side. Most often, storms move northward or curve to the northeast once in the middle latitudes, with the heaviest rain on the west side and strongest winds to the east. This was the situation that led to excessive rainfall and record river flooding with the coastal tracks of Hurricane Floyd in 1999 and Hurricane Irene in 2011. Sandy's unusual east-to-west path across southern New Jersey produced an unusual distribution of precipitation and wind.

Surge

It was the unusual path of Sandy that resulted in the extreme coastal storm surge along central and northern coasts of New Jersey and the New York Harbor as well as in adjacent bays, harbors, rivers, and wetlands.[12] Ocean levels are measured as the height above the mean lower low water, a metric reflecting low tide influences. The highest level, a measurement of 14.4 feet above mean lower low water at coastal Sandy Hook during the evening of October 29, was 4.3 feet above the previous century record established during Hurricane Donna in September 1960. This level was approximately nine feet above what would normally be expected at the coincident high tide. Contributing to the surge was the previously mentioned long fetch of wind toward the coast and the fact that landfall came at high tide and under a full moon. Also, sea level is presently approximately one foot higher than a century earlier, owing to warmer water (thermal expansion), glacial ice melting, and geologic subsidence along the New Jersey coast.

A CHANGE IN TROPICAL WARNING PROCEDURE

As Sandy traveled up the East Coast, the National Hurricane Center forecast a transition from a hurricane to an extratropical storm prior to landfall. Thus the National Weather Service was concerned that continuing hurricane warnings north of Cape Hatteras, North Carolina, might confuse the public once the transition took place, so none were ever issued. Also, once a storm was no longer deemed tropical, forecasting responsibility traditionally was turned over from the National Hurricane Center to local NWS offices. Although this handoff from the hurricane center to the local weather service office allows the hurricane center to focus on its specialty of tracking tropical storms, it may also inadvertently suggest to the public that the storm's intensity has diminished.

Ensuing debate after Sandy included claims that the lack of hurricane or tropical storm warnings along the mid-Atlantic coast may have lulled decision makers and the public into a false sense of security. This led to a decision by the NWS to broaden the criteria for tropical storm watches and warnings to allow its National Hurricane Center to continue tracking large tropical storms and issuing advisories even after they no longer technically have tropical characteristics, if there is a major potential for danger to life or property. This policy took effect with the 2013 Atlantic hurricane season.[13]

CONCLUSION

As devastating and deadly as Sandy was, there have been worse storms in the mid-Atlantic region. This includes the aforementioned 1938 hurricane; a hurricane in 1821 that made landfall in Cape May, with sustained winds of perhaps more than 100 mph; and a hurricane in 1944 that remained just off the New Jersey coast but brought heavy rain, strong winds, and a significant surge. With a warming atmosphere and ocean and a resultant increase in atmospheric moisture, the climate system is increasingly primed for wetter, more intense storms.[14] Of course, there must be atmospheric disturbances, triggers if you will, to generate such destructive events. However, with sea levels expected to rise another foot by midcentury and three feet by 2100, even lesser storms that would have done little harm in previous years may be quite damaging.

That Sandy arrived a year to the day from a record-shattering October 2011 snowstorm in central and northern New Jersey adds a sobering and quite remarkable footnote to the events surrounding October 29, 2012.[15] The size and strength of Sandy was the result of a unique combination of atmospheric and oceanic conditions. Sandy's storm surge far surpassed the previous record of the past century; inland winds had never been as strong or of a multi-hour duration in the modern era across central and northern New Jersey; and rainfall in far southern New Jersey had a return period of approximately two hundred years. This resulted in storm damage that equaled or exceeded the worst coastal and inland battering of the past century or longer in many portions of the state. Tragically, the New Jersey death toll of approximately forty from direct or indirect effects of the storm made this the deadliest natural event on record in New Jersey.[16] Some speculation exists among atmospheric scientists and critical decision makers as to whether such a rogue storm as Sandy may become more frequent as climate change continues.[17]

LESSONS

Hurricane Sandy delivered several lessons about the potential effects of storms in the densely populated New York–New Jersey region that apply to other urbanized coastal areas along the eastern and Gulf coasts:

- Tropical and post-tropical cyclones can make landfall in the mid-Atlantic from the southeast.

New Jersey and many other areas along the coast usually experience only the indirect effects of tropical storms, but direct hits happen periodically and can cause massive damage.

- Under the right conditions, large amounts of water can be funneled into the bays and harbors, leading to significant storm surges in the surrounding coastal regions.

Waters from storm surge damage during Sandy occurred in areas much farther inland than residents and authorities expected. Bays and harbors all along the eastern and Gulf coasts are vulnerable in the same way.

- Sandy was not the worst storm possible in the region.

Storm damage was due primarily to wind and surge damage, not rainfall. A storm that moved more slowly over the region could have caused much more coastal and inland flooding damage than Sandy did. And winds could certainly be stronger along the coast and inland during future, potentially worse, storms.

- Allowing the National Hurricane Center to continue issuing forecasts and warnings for a storm that is no longer technically tropical may improve communication to the public.

Weather services in the United States and elsewhere are accustomed to adjusting their forecasting and communication procedures after storms. Climate change and changes in storm behavior will likely challenge forecasters' communication efforts even more in the near future.

NOTES

1. National Oceanic and Atmospheric Administration (NOAA), *Hurricane/Post-Tropical Cyclone Sandy October 22–29, 2012: Service Assessment* (NOAA/National Weather Service (NWS), 2013), accessed June 6, 2015, http://www.nws.noaa.gov/os/assessments/pdfs/Sandy13.pdf.
2. Eric S. Blake, Todd B. Kimberlain, Robert J. Berg, John P. Cangialosi, and John L. Beven II, *Tropical Cyclone Report: Hurricane Sandy*, Technical Report AL182012 (NOAA/National Hurricane Center, 2013), accessed June 18, 2014, http://www.nhc.noaa.gov/data/tcr/AL182012_Sandy.pdf.
3. *Landfall* refers to the time at which the center of a storm's circulation (i.e., its eye) crosses a coastline.
4. *Sustained winds* are determined by averaging the instantaneous wind speed over some time interval. Different time intervals can be used. Here, the best track data uses a one-minute interval, consistent with National Hurricane Center policy. However, the National Weather Service uses a two-minute interval, and most observations (including those described later in this chapter) follow the NWS definition.
5. Michael C. Morgan and John W. Nielsen-Gammon, "Using Tropopause Maps to Diagnose Midlatitude Weather Systems," *Monthly Weather Review* 126, no. 10 (1998): 2,555–2,579.

6. Kyle S. Mattingly, Jordan T. McLeod, John A. Knox, J. Marshall Shepherd, and Thomas L. Mote, "A Climatological Assessment of Greenland Blocking Conditions Associated with the Track of Hurricane Sandy and Historical North Atlantic Hurricanes," *International Journal of Climatology* 35, no. 5 (2015): 746–760.
7. Linus Magnusson, Jean-Raymond Bidlot, Simon T. K. Lang, Alan Thorpe, Nils Wedi, and Munehiko Yamaguchi, "Evaluation of Medium-Range Forecasts for Hurricane Sandy," *Monthly Weather Review* 142, no. 5 (2014): 1,962–1,981.
8. Thomas J. Galarneau Jr., Christopher A. Davis, and Melvyn A. Shapiro, "Intensification of Hurricane Sandy (2012) through Extratropical Warm Core Seclusion," *Monthly Weather Review* 141, no. 12 (2013): 4,296–4,321.
9. Thomas M. Hamill, Jeffrey S. Whitaker, Michael Fiorino, and Stanley G. Benjamin, "Global Ensemble Predictions of 2009's Tropical Cyclones Initialized with an Ensemble Kalman Filter," *Monthly Weather Review* 139, no. 2 (2011): 668–688.
10. Magnusson et al., "Evaluation of Medium-Range Forecasts."
11. Louis Uccellini, "Sandy—One Year Later," National Weather Service, October 28, 2013, http://www.nws.noaa.gov/com/weatherreadynation/news/131028_sandy.html.
12. Timothy M. Hall and Adam H. Sobel, "On the Impact Angle of Hurricane Sandy's New Jersey Landfall," *Geophysical Research Letters* 40, no. 10 (2013): 2,312–2,315.
13. NWS, *Change in Tropical Cyclone Watch and Warning Definitions to Include Post-Tropical Cyclones Effective June 1, 2013*, Service Change Notice 13-28 (NOAA/National Weather Service, 2013), accessed June 6, 2015, http://www.nws.noaa.gov/om/notification/scn13–28tropical_watch-warn.htm.
14. Thomas R. Knutson, John L. McBride, Johnny Chan, Kerry Emanuel, Greg Holland, Chris Landsea, Isaac Held, James P. Kossin, A. K. Srivastava, and Masato Sugi, "Tropical Cyclones and Climate Change," *Nature Geoscience* 3, no. 3 (2010): 157–163.
15. Douglas LeComte, "U.S. Weather Highlights 2011: Unparalleled Weather Extremes," *Weatherwise* 65, no. 3 (2012): 20–27.
16. Blake et al., *Tropical Cyclone Report*, 14.
17. Jennifer A. Francis and Stephen J. Vavrus, "Evidence Linking Arctic Amplification to Extreme Weather in Mid-latitudes," *Geophysical Research Letters* 39, no. 6 (2012): L06801, doi: 10.1029/2012GL051000.

2 • A TOUGH MOVE TO MAKE

Lessons Learned from Emergency Evacuations in Coastal Connecticut during Hurricane Sandy

DANIEL BALDWIN HESS
AND BRIAN W. CONLEY

With an increasing global population and a changing climate, more people are susceptible to hazards, both natural and human-made, than ever before. This chapter explores how emergency managers cope with the complexity and chaos of evacuations under a changing climate by investigating evacuations that occurred in coastal Connecticut in response to Hurricane Sandy in October 2012. We identify key lessons, transferrable to other locations and emergency situations, which can be applied by emergency managers to better prepare for and effectively coordinate large-scale evacuations.

AN UNPRECEDENTED STORM

Decisions about evacuations in Connecticut began early. On October 22, 2012, a circulating low-pressure system grew in power as it moved through the Caribbean Sea, eventually becoming a tropical storm dubbed Sandy. On October 25, when it became a Category 3 storm with sustained winds of

115 miles per hour and made its second landfall over Cuba, the state of Connecticut began preparing for the storm, a full day before the U.S. National Weather Service issued tropical storm warnings for parts of Florida.

On October 27, while the outward reaches of the storm swept over the North Carolina coast, Connecticut Governor Dan Malloy signed a Declaration of Civil Preparedness for the state. President Barack Obama issued preemptive emergency declarations for several states, including Connecticut, authorizing direct federal support to the preparations under way. Hurricane Sandy made landfall in the United States in southeastern New Jersey on the evening of October 29, 2012, extending high winds and storm surge to Connecticut and other nearby states. On October 30, just one day after impact, presidential Disaster Declarations were issued for New Jersey, New York and Connecticut.[1]

Local and state emergency managers faced critical decisions about whether or not to order mandatory evacuations, when to make an order effective, and which neighborhoods to include. Hundreds of thousands of citizens in low-lying areas from Virginia to Rhode Island removed themselves from danger, most in response to mandatory evacuation orders.[2] Managers know that evacuation expends the region's emergency capacity and requires individuals to disrupt their lives and separate themselves from their property, possessions, and loved ones.[3] The benefits of undertaking an evacuation must therefore surpass, with a high degree of certainty, the costs (in terms of property damage, injury, and loss of life) of forgoing an evacuation. This calculation occurs at two levels: government decision makers ordering a mandatory evacuation; and individuals deciding to vacate, whether or not a mandatory evacuation has been ordered. By reviewing events in the state of Connecticut during Hurricane Sandy, we provide insight on current and best practices in the management of a large-scale evacuation by state and local governments.

RESEARCH PLAN

Scholarly research on large-scale evacuation forms the foundation of this inquiry. We review documents from municipal and state emergency planning departments, including disaster plans and government agencies' After Action Reports (AAR), which evaluate disaster response and summarize lessons learned. We also review news reporting before, during, and after Hurricane Sandy and analyze geographic information on evacuations. We report on one-on-one interviews with more than a dozen local emergency planning chiefs and state representatives and with disaster experts from academia and practice. This chapter thus extends research in evacuation policy and planning, disaster response, and studies of people who cannot self-evacuate.[4]

We set the stage by examining the factors that play a key role in the evacuation decision at two scales, individual and government decision makers, and also by reviewing the prevailing policies currently guiding large-scale evacuations in the Connecticut and the United States.

Individuals' Decisions to Evacuate

A person's willingness or ability to vacate an area with necessary urgency can be limited, as demonstrated during Hurricane Katrina, when some individuals who did not evacuate required dangerous rescue by emergency personnel after floodwaters rose.[5] Even when people are properly informed about disaster risk, it cannot be assumed that the provision of a warning will provide individuals with a sound appreciation of the implications of that warning, or that awareness will lead people to take action.[6] Individuals will pursue the action that they deem to be most sensible, based on the information provided and their own perspective.[7]

The most fundamental predictor of evacuation participation is an individual's knowledge of an evacuation order.[8] But even when an official order is placed, it has been estimated that one-third of the public will not necessarily comply.[9] Individuals' and families' reasons for not complying with mandatory evacuation orders are complex.[10] People are more likely to comply with evacuation orders if the procedure for evacuation is clearly explained and if they judge the information source to be credible, for example, from a known government official.[11] The risk level of a specific location, along with residents' perception of this risk, both play a part in the likelihood of evacuating.[12] News and weather media prompt increasing levels of evacuation participation by presenting footage that can sometimes be sensational.[13] However, during many emergency incidents, an evacuation order may become muddled, resulting in individuals ignoring the order.[14] The biggest impediment to evacuation may be conflicting evacuation orders or evacuation orders that conflict with other information given by authorities.[15] For example, contradictory estimates of risk and ineffective critical communications between higher levels of government contributed to New Orleans Mayor Ray Nagin's delay in activating a formal evacuation order before Hurricane Katrina in 2005, which came less than a day before the storm made landfall and was one factor in the high number of fatalities caused by the storm.[16]

An individual's life experiences, especially those attached to previous emergencies, are generally the most accepted factor influencing a decision to evacuate.[17] Numerous studies consider whether prior disaster experience affects an individual's evacuation decision, but findings are contradictory.[18] One study found more specifically that people who survive a previous hurricane without evacuating are less likely to comply with evacuation orders for later storms.[19]

People over seventy years of age have been found less likely than younger groups to obey evacuation orders, while households with children are more likely to follow evacuation orders.[20] Residing in mobile housing, having a high-level of educational attainment, or having a high income also increases household members' willingness to evacuate.[21] Owning a vehicle greatly influences the ability and willingness to evacuate.

Research supports the premise that mandated evacuations do trigger higher evacuation participation rates than voluntary orders before hurricanes.[22] But this may not hold true under all scenarios, as certain factors, such as skepticism of top-down directives and mistrust of government information regarding disasters, have led to unfavorable perceptions of mandated evacuations among affected individuals interviewed after disasters.[23]

Individuals' perceptions of vulnerability affect evacuation response. Those who feel their risk to be high based on proximity to a coast decide to evacuate their homes more quickly but also spend more time securing homes and property, likely because of their perceived higher level of risk, than those living further inland. So, without explicit direction to behave otherwise, those living further inland with a lower degree of risk, will evacuate at essentially the same time as coastal residents, which could lengthen overall evacuation times for everybody.[24] One study of New Jersey during the evacuation for Hurricane Irene in 2011 showed that limited prior experience with hurricane evacuations, a high concentration of tourists, and evacuation routes of limited capacity quicken a population's response to an evacuation order.[25] However, analyses of behavior, such as the role of leadership in evacuations, are perhaps better metrics for understanding evacuation decision making.[26]

Government Decisions to Mandate Evacuation

Local government leaders are best poised to prevent or reduce the local effects of a disaster and to decide what resources, if any, are to be requested from higher levels of government to carry out the evacuation.[27] The clarity of each municipality's emergency plan and the ability to implement that plan vary by jurisdiction and emergency situation.

Estimating risk is a difficult task, given the often incomplete or inaccurate predictions of the severity or extent of extreme weather events.[28] Additionally, city managers often misperceive vulnerabilities in their communities to certain types of disasters.[29] With limited political incentive for local decision makers to seek funding for emergency preparedness programs, preparedness has often been inadequate, even though the loss of life and property in recent disasters has underscored this need.[30]

As news of a storm or other threatening event arrives, emergency managers face a dilemma. Forgoing an evacuation may result in many being injured or killed, but ordering an evacuation unnecessarily may create its own dangers. In some evacuations, shadow evacuations, when a sizable share of evacuees reside outside of an official evacuation zone, can impede the evacuation of those who are most vulnerable.[31] If local officials extend evacuation directives to areas that are not at a high level of risk, they can exacerbate the shadow evacuation problem. More spatially explicit evacuation orders, such as those based on the National Oceanographic and Atmospheric Administration's SLOSH maps (Sea, Lake, and Overland Surges from Hurricanes) can reduce uncertainty about evacuation. Officials can also improve their designations by using decision support systems such as checklists or decision trees.[32]

Other governmental and nongovernmental institutions also supply critical intelligence for decision making. Natural hazards lead to evacuations more commonly than other types of emergencies, and media can play a role in government officials' decision to evacuate from weather events.[33] Information provided by the National Weather Service and other media during an extreme weather event influences how individuals decide about evacuating, whether or not a mandatory evacuation is ordered.[34] The condition of critical infrastructure systems and the special needs of certain populations—for example children, elderly, and handicapped—are also key considerations in deciding to initiate a large-scale evacuation.[35]

If a large-scale evacuation is not warranted, emergency managers also reserve the option to keep and provide for affected populations in a disaster zone, a strategy known as shelter-in-place, which is thought to be safer under certain conditions than evacuating.[36] In a review of shelter-in-place decisions related to wildfires, several broad sets of factors were determined to play a role in individuals' decision to evacuate or remain in place, including hazard level, community context, warning and evacuation factors, and policy context, understood as the political, regulatory, and legal framework that shapes evacuation decisions made by emergency managers and elected officials.[37] Although threats differ for fires versus storms, these factors can also be considered for other hazards, including tropical storms. Other studies have found that sheltering-in-place may be the best option when the duration of an event is extremely short or when an evacuation would unnecessarily expose people to known danger, for example, if there is contamination along emergency evacuation routes.[38]

Critical Factors in the Decision to Evacuate

The decision to undertake an evacuation arises as a wicked problem, a decision that most often cannot be made by using a replicable, systematic formula.[39] However, several factors are understood to influence whether or not an evacuation should, or will, occur. These include (a) the time available for officials to designate a storm or other threat as an event warranting an evacuation notice; (b) the severity, nature, and duration of disaster effects, including weather and climate conditions; (c) the resources available for sheltering; (d) the condition of critical infrastructure, such as roads that may be flooded; (e) the community context, that is, the preparedness and training of individuals and demographic composition of the community; and (f) the policy context, which includes the preparedness and training of local government personnel.[40]

CONNECTICUT'S EXPERIENCE WITH SANDY

Lessons from evacuation research are often circulated among emergency professionals, although putting such lessons into practice is difficult. The federal structures of the United States place a heavy burden on municipalities for impending hazard response and emergency evacuations. Connecticut's municipalities provide a case study of orderly coordination based on training and experience in a state that received considerable damage but that was not the most directly hit.

Evacuation Policy and Protocol

Disaster response and evacuation protocol in the United States can generally be viewed as bottom-up, placing authority for disaster response principally on an affected jurisdiction in ascending order: village, town, city, county, and state.[41] Most states, including Connecticut, place the decision to evacuate a locale with the chief executive officer of each municipality, if not the governor of the state.[42] State governments, especially governors, are typically afforded comparable authority to issue an evacuation order.[43] The federal government assists local and state governments when their capacity to respond is exceeded, especially evacuations involving the movement of people across state lines.[44]

This means that primary legal authority during disaster rests with state governments.[45] The Federal Emergency Management Administration (FEMA) provides an exemplar Local Emergency Plan, which the State of Connecticut has recommended in adapted form for its municipalities. The four key roles in this plan include (1) the municipal executive, who has the authority to issue an evacuation order; (2) a public information officer, tasked with communicating evacuation orders to the public; (3) an evacuation coordinator, to determine evacuation routes and the process to return people home after the evacuation period; and (4) the police, to help alert the public, control traffic, and secure the area.

In 2004, Connecticut passed Public Act 11-51, creating the Division of Emergency Management and Homeland Security (DEMHS) to more effectively execute emergency management, civil preparedness, and other homeland security missions by enhancing collaboration between municipal and state agencies. Defining a coordinated and proactive regional approach to preparedness, this act divided Connecticut into five emergency management regions and required Regional Emergency Planning Teams (REPT) for each region. This policy facilitates municipal communication regarding needs and allows municipalities to share resources and reduce costs.

Safeguarding Connecticut for Hurricane Sandy: How, When, and Where to Evacuate

The first tidal surge in Connecticut began on Sunday, October 28, but preliminary evacuations began on Friday, October 26, forty-eight hours earlier, according to protocol. Evacuations continued in earnest on Saturday, one day before landfall. To publicize the impending storm, the need for preparation, and, in some communities, the order to evacuate, the State of Connecticut emergency notification system was used, along with local code red emergency notification systems in various municipalities. Local emergency planners in more than twenty-five municipalities ordered evacuations of threatened areas in response to the direction from the governor of Connecticut to zones marked as most vulnerable in the SLOSH model maps prepared by the National Hurricane Center. As shown in figure 2.1, these evacuation zones included many populous areas of Connecticut's southern coast along the Long Island Sound, including the roughly fourteen towns in the southwestern portion that were hardest hit in the state by both Hurricanes Irene and Sandy.[46] The Town of Fairfield experienced the most destruction in both storms.

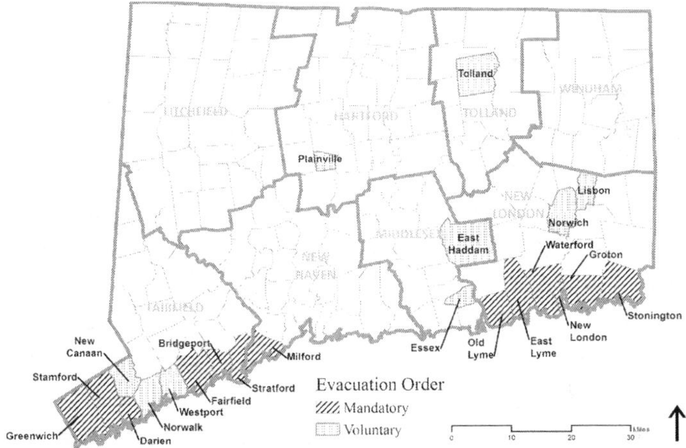

FIGURE 2.1. Towns issuing evacuation orders for Hurricane Sandy in Connecticut. Evacuation orders were issued only for areas lying within Category 4 SLOSH zones.

It is difficult to determine exactly how many people in Connecticut evacuated their homes for Hurricane Sandy. In Bridgeport, the largest city in the state, officials visited approximately thirteen thousand individual households to deliver evacuation notices. Officials there eventually provided buses to evacuate six hundred students from the University of Bridgeport.[47] In Westport, the flood zone encompassed about three hundred properties and more than five hundred individuals, according to local emergency managers. In the Town of Greenwich, approximately nine hundred people resided in a mandatory evacuation zone,[48] and in the City of Norwalk, more than two thousand people resided in zones where evacuation was recommended but not mandatory. In some municipalities, officials were especially concerned about public housing complexes, senior housing, and other residential institutions that were in evacuation zones.

Local governments generally took responsibility for evacuating disaster zones, overseeing warming stations or shelters that were operated by the Red Cross, local Medical Reserve Corps, or other groups, and offering assistance during utility outages.[49] The statutes of some municipalities in Connecticut allow for mandatory evacuations, while in other municipalities, evacuation guidance is ordered but it is not mandatory. In a mandatory evacuation, a person not complying with the order is considered to be breaking the law, and he or she can be arrested, although emergency managers interviewed for this project pointed out that arrests are seldom made, which weakens the meaning of a mandatory order. In response to Hurricane Sandy, thirteen Connecticut towns issued mandatory evacuations, including most of the impacted towns along the coast. Three towns in the heavily impacted southwestern section of Connecticut, New Canaan, Norwalk, and Westport, were among the towns issuing voluntary evacuation orders. Municipalities' vulnerability to surge from hurricanes is shown in figure 2.2.[50]

Many emergency planners we interviewed reported that police, fire departments, and emergency management staff began driving through neighborhoods two days prior to the first storm surge striking the Connecticut coast, making announcements through public address systems and visiting each home in evacuation zones, telling residents about the impending storm and hand delivering a written evacuation notice. As in every evacuation, some residents told emergency workers that they would not evacuate. Some towns required these people to sign a waiver acknowledging that they

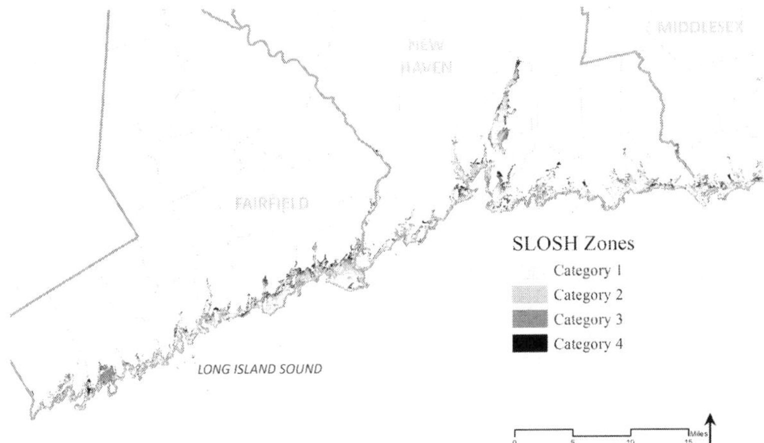

FIGURE 2.2. Sea, lake, and overland surges from hurricanes (SLOSH) zones in southwestern Connecticut. All shaded areas were evacuated during Hurricane Sandy.

had been provided with a written evacuation order, aimed at reducing the liability of the municipality if people suffered personal injury or property damage. In towns where mandatory evacuation was ordered, emergency workers informed people who said they would not likely evacuate that the municipality would remove children, to protect their safety, and suspend utility service, to increase the safety of emergency personnel.

Some people who chose not to evacuate later changed their minds and placed distress calls for assisted evacuation. This is exactly what emergency managers attempt to avoid, because of the dangers posed to citizens and emergency workers by urgent evacuation, which often happens in deteriorating conditions. In the City of Norwalk, high-water vehicles were used to rescue more than one hundred people from homes and porches shortly before the storm. Across the state, more than 144 emergency rescues were carried out by local first responders, and additional search-and-rescue missions were conducted by the National Guard.[51]

Most Connecticut communities in evacuation zones included mixed-income residential areas. Many families had the means to easily self-evacuate by automobile, which is the assumed default action by the sample town evacuation plan used in Connecticut. Emergency managers that we interviewed, however, mentioned that there were carless populations in evacuation zones in all communities. Emergency planning efforts had preidentified the general location of carless people and coordinated with local transit agencies to move people from residential neighborhoods to shelters. This generally worked well according to emergency managers, with no significant problems reported. Some towns distributed flyers, in English and Spanish, during door-to-door visits that included information about shelters and public transportation options for getting there, including preplaced vehicles at staging points near known vulnerable populations. The Norwalk Transit District and other transit agencies provided such service free of charge for passengers.

LESSONS

The devastating impact of Sandy on communities throughout Connecticut is undeniable. As a result of the storm, hundreds of local roads and two hundred state routes became impassable, 650,000 electric customers lost power, and six people lost their lives (none of which were related to evacuation).[52] One year after the storm, FEMA had approved about $316 million to assist households and small businesses throughout the state to recover from the storm.[53] However, the potential impact on individuals in Connecticut was considerably lessened by the proactive and coordinated evacuations that took place across the state. Interviews with emergency officials in Connecticut and review of state and local After Action Reports, reveal several lessons for emergency managers in the United States and around the world who are working toward community resilience in the face of climate and other potential hazard events. Pivotal key factors include the following:

- Officials' preparation based on recent disaster experience is essential for easing later evacuations.

Three significant emergency events impacted Connecticut in the space of a little more than one year. First, Hurricane Irene (August 2011); second, a statewide hurricane exercise that was part of the governor's Emergency Planning and Preparedness Initiative (July 2012); and third, Hurricane Sandy (October 2012). All emergency managers that we interviewed mentioned that readiness for Hurricane Sandy was improved by this sequence of events. Each disaster allowed emergency workers the opportunity to enact emergency operations, assess outcomes, and make improvements in responses to the variety of storms that Connecticut can experience: blizzards, hurricanes, and nor'easters, which usually come with advance notice and which require similar preparation, response, and recovery.

In recent disasters, emergency managers had learned that evacuees will arrive at shelters, get settled, and then make phone calls to family and friends. After making contact, many will get picked up by family and friends and stay elsewhere, resulting in fewer people officially sheltered when a storm hits. If a shelter opens later, fewer arriving people will have time to make other arrangements before the storm hits. In Hurricane Sandy, some shelters were set up and not used, but at least 3,534 people sheltered overnight at official locations in Connecticut,[54] as shown in figure 2.3.[55]

- Reviewing miscommunication from past evacuations can improve how officials share information with residents and increase evacuation compliance.

In interviews, emergency managers in Connecticut stressed the importance of clear and timely information, delivered in a variety of formats, with door-to-door visits now considered to be crucial for raising compliance rates and reducing municipalities' liability. One example of learning about

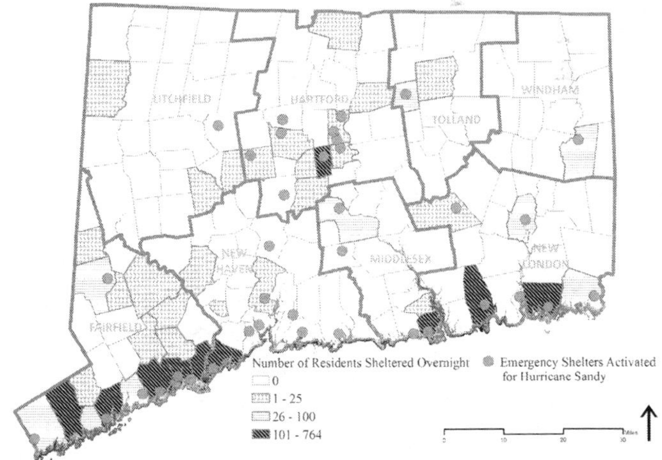

FIGURE 2.3. Municipal shelters opened in Connecticut during Hurricane Sandy and the number of residents who were sheltered overnight.

clarity came from the City of Norwalk, which, before Hurricane Irene, had released a hurricane map with at least four colors showing zones with different risk levels. City workers were flooded with calls from people asking for clarification about whether they needed to evacuate. Before Hurricane Sandy, managers released a simpler map with one evacuation zone, colored pink, and this proved to be beneficial.

- Coordinated government efforts can encourage communities to increase their resilience after disasters.

The state of Connecticut has sound guidelines for community evacuation, which the emergency management staff of the municipalities generally followed. These guidelines and other proactive measures have enhanced the resilience or, "the ability of a social system to respond and recover from disasters and . . . allow the system to absorb impacts and cope with an event" in coastal communities of Connecticut.[56] Information sharing can be one way of improving a community's resilience. In Connecticut, local emergency management plans are created and made available to every resident, identifying leadership, detailed emergency planning, and the relevant local hazards. Municipalities, with state encouragement, also endeavored to improve their resilience to potential disasters and to hurricanes in particular, such as trimming and cutting trees to improve safety during emergencies and reduce power outages, which was carried out with the assistance of a State Vegetation Management Task Force. Through regional coordination for emergency management and development of Regional Emergency Planning Teams, the state created a process by which adjacent towns can share resources when reasonable, such as sharing shelters depending on demand, power availability, and proximity to an evacuation zone.

Additional benefits of coordination and unification of functions have been realized as towns have reduced or removed duplication of efforts by improving their Emergency Operations Centers (EOCs), which can act as "one-stop shops" for municipality-wide reporting during emergencies. Until recently, the police, fire, and public works departments all fielded and responded to calls during an emergency. Now, each department completes an electronic report when a complaint arises, and EOC workers organize the reports and assign a single response to an appropriate agency.

Relationship building can strengthen the ability of an emergency planning department to do superior work. In interviews, emergency management leaders mentioned their work to develop strong ties with public transit agencies, public housing authorities, public works departments, and other public entities concerning emergency planning, response, and recovery.

- Local emergency managers who actively seek information can gain insight from the experiences of other communities.

Emergency managers reported that they learned much from the disaster management experience in New Orleans with Hurricane Katrina by reading documents, attending national hurricane conferences, and observing policy and operational shifts in affected states. Practical information from southern states about hurricane experience was useful and was put into practice for the emergency response and mobilization of resources in Connecticut for Hurricane Irene and Hurricane Sandy.

FUTURE NEEDS

This research validates that there is an emergent paradigm in disaster response and evacuation centered on proactive planning and collaboration between government entities, organizations, and community members. And though the practice of large-scale evacuations in response to oncoming tropical storms has improved over the past decade, responses must be evaluated and improved after every actual evacuation.

An important concern is the need to expand emergency training exercises with private sector partners, such as drills for restoring telecommunications. Local emergency planners also mentioned their wish to collaborate with FEMA to improve hazard mitigation and to develop better long-term recovery plans. For example, local detailed inventories of all dwellings and their elevations would inform residents and community planners about potential storm hazards.

Hurricane Sandy did not profoundly transform the processes and human dynamics involved in large-scale evacuations in the state of Connecticut. Rather, Sandy can be seen as an important event in the continuing advancement of large-scale evacuation management, a supposition that may not be transferrable to all regions of the United States, or the world. The events in Connecticut during Hurricane Sandy do prove that lessons have been learned there from previous storms and will be learned from Sandy.

NOTES

This research was partially funded by a grant through the U.S.D.O.T./RITA through the University Transportation Research Center–Region II. The authors acknowledge invaluable contributions to the research from Scott Appleby, Brenda Bergeron, Michele DeLuca, Amy Donahue, Syma Ebbin, Christina Farrell, Andy Kingsbury, Robert Koch, Fran Raiola, Kimberly Schueler, and Dan Warzoha.

1. Federal Emergency Management Agency (FEMA), "Hurricane Sandy: Timeline," accessed February 2, 2014, https://www.fema.gov/media-library/assets/documents/31987.
2. John H. Cushman, "Coastal Surge Forecast from Hurricane Sandy Prompts Evacuations," *New York Times*, October 27, 2012.
3. Amy L. Fairchild, James Colgrove, and Marian Moser Jones, "The Challenge of Mandatory Evacuation: Providing For and Deciding For," *Health Affairs* 25, no. 4 (2006): 958–967.
4. Daniel B. Hess, Brian Conley, and Christina Farrell, "Improving Transportation Resource Coordination for Multi-modal Evacuation Planning: A Literature Review and Research Agenda," *Journal of the Transportation Research Board* 2376, no. 1 (2014); Hess, Conley, and Farrell, *Barriers to Resource Coordination for Multi-modal Evacuation Planning* (Buffalo, NY: Multidisciplinary Center for Earthquake Engineering Research, University at Buffalo, 2013); Daniel B. Hess and Lucy A. Arendt, *Enhancements to Hospital Resiliency: Improving Emergency Planning for and Response to Hurricanes* (Buffalo, NY: Multidisciplinary Center for Earthquake Engineering Research, University at Buffalo 2006); Hess and Arendt, *Enhancements to Hospital Resiliency: Improving Emergency Planning for and Response to Hurricanes: MCEER Technical Report 09-0007* (Buffalo, NY: Multidisciplinary Center for Earthquake Engineering Research, University at Buffalo, 2009); Daniel B. Hess and Julie C. Gotham, "Multi-modal Mass Evacuation in Upstate New York: A Review of Disaster Plans," *Journal of Homeland Security and Emergency Management* 4, no. 3, art. 11 (2007): 1–19; John L. Renne, Thomas W. Sanchez, Pam Jenkins, and Robert Peterson, "Challenge of Evacuating the Carless in Five Major US Cities," *Transportation Research Record: Journal of the Transportation Research Board* 2119, no. 1 (2009): 36–44.
5. Michael K. Lindell, Jing-Chein Lu, and Carla S. Prater, "Household Decision Making and Evacuation in Response to Hurricane Lili," *Natural Hazards Review* 6, no. 4 (2005): 171–179; Keith Elder et al., "African Americans' Decisions Not to Evacuate New Orleans before Hurricane Katrina: A Qualitative Study," *American Journal of Public Health* 97

(2007); Hess and Arendt, *Enhancements to Hospital Resiliency: Improving Emergency Planning for and Response to Hurricanes* (2006); Hess and Arendt, *Enhancements to Hospital Resiliency: Improving Emergency Planning for and Response to Hurricanes* (2009).

6. John H. Sims and Duane D. Baumann, "Educational Programs and Human Response to Natural Hazards," *Environment and Behavior* 15, no. 2 (1983): 165–189.

7. John H. Sorensen, "When Shall We Leave? Factors Affecting the Timing of Evacuation Departures," *International Journal of Mass Emergencies and Disasters* 9, no. 2 (1991): 153–165.

8. Earl J. Baker, "Hurricane Evacuation Behavior," *International Journal of Mass Emergencies and Disasters* 9, no. 2 (1991): 287–310; Samiul Hasan et al., "Behavioral Model to Understand Household-Level Hurricane Evacuation Decision Making," *Journal of Transportation Engineering* 137, no. 5 (2010): 341–348.

9. Michael T. Carter, *Community Warning Systems: Interface between the Broadcast Media, Emergency Service Agencies, and the National Weather Service*, Minneapolis: University of Minnesota, Department of Sociology, 1979.

10. Linda I. Gibbs and Caswell F. Holloway, *Hurricane Sandy after Action: Report and Recommendations to Mayor Michael R. Bloomberg* (New York: City of New York, 2013).

11. Sims and Baumann, "Educational Programs and Human Response to Natural Hazards."

12. Baker, "Hurricane Evacuation Behavior."

13. Sims and Baumann, "Educational Programs and Human Response to Natural Hazards"; Elder et al., "African Americans' Decisions Not to Evacuate New Orleans before Hurricane Katrina."

14. Sims and Baumann, "Educational Programs and Human Response to Natural Hazards"; Baker, "Hurricane Evacuation Behavior"; Sorensen, "When Shall We Leave?"; Daniel Baldwin Hess and Christina M. Farrell, "Evacuation from Disaster Zones: Lessons from Recent Disasters in Australia and Japan," in *Handbook on Securing Transportation Systems*, ed. Simon Hakim and Yoram Shiftan (London: John Wiley and Sons, forthcoming).

15. Elder et al., "African Americans' Decisions Not to Evacuate New Orleans before Hurricane Katrina."

16. David S. Heller, "Evacuation Planning in the Aftermath of Katrina: Lessons Learned," *Risk, Hazards & Crisis in Public Policy* 1, no. 2 (2010): 131–174.

17. Sims and Baumann, "Educational Programs and Human Response to Natural Hazards"; Susan L. Cutter and Mark M. Smith, "Fleeing from the Hurricane's Wrath: Evacuation and the Two Americas," *Environment: Science and Policy for Sustainable Development* 51, no. 2 (2009): 26–36; Jasmin K. Riad, Fran H. Norris, and R. Barry Ruback, "Predicting Evacuation in Two Major Disasters: Risk Perception, Social Influence, and Access to Resources," *Journal of Applied Social Psychology* 29, no. 5 (1999): 918–934.

18. Sims and Baumann, "Educational Programs and Human Response to Natural Hazards"; Lindell, Lu, and Prater, "Household Decision Making and Evacuation in Response to Hurricane Lili"; Elder et al., "African Americans' Decisions Not to Evacuate New Orleans before Hurricane Katrina."

19. Baker, "Hurricane Evacuation Behavior."

20. Harun Rasid, Wolfgang Haider, and Len Hunt, "Post-flood Assessment of Emergency Evacuation Policies in the Red River Basin, Southern Manitoba," *Canadian Geographer/Le Géographe Canadien* 44, no. 4 (2000): 369–386; Hasan et al., "Behavioral Model to Understand Household-Level Hurricane Evacuation Decision Making."

21. Hasan et al., "Behavioral Model to Understand Household-Level Hurricane Evacuation Decision Making."

22. Ibid.

23. Rasid, Haider, and Hunt, "Post-flood Assessment of Emergency Evacuation Policies in the Red River Basin, Southern Manitoba."

24. Lindell, Lu, and Prater, "Household Decision Making and Evacuation in Response to Hurricane Lili."

25. Jian Li et al., "Empirical Evacuation Response Curve during Hurricane Irene in Cape May County, New Jersey," *Transportation Research Record: Journal of the Transportation Research Board* 2376, no. 1 (2013): 1–10.

26. Benigno E. Aguirre, "Emergency Evacuations, Panic, and Social Psychology," *Psychiatry: Interpersonal and Biological Processes* 68, no. 2 (2005): 121–129.

27. Scott Somers and James H. Svara, "Assessing and Managing Environmental Risk: Connecting Local Government Management with Emergency Management," *Public Administration Review* 69, no. 2 (2009): 181–193.

28. FEMA, "National Association of State Departments of Agriculture Emergency Response Plan: Mass Evacuation Incident Annex," accessed January 4, 2012, http://www.fema.gov/pdf/emergency/nrf/nrf_massevacuationincidentannex.pdf.

29. Dianne Rahm and Christopher G. Reddick, "US City Managers' Perceptions of Disaster Risks: Consequences for Urban Emergency Management," *Journal of Contingencies and Crisis Management* 19, no. 3 (2011): 136–146.

30. Robert P. Wolensky and Kenneth C. Wolensky, "Local Government's Problem with Disaster Management: A Literature Review and Structural Analysis," *Review of Policy Research* 9, no. 4 (1990): 703–725; Somers and Svara, "Assessing and Managing Environmental Risk."

31. Fairchild, Colgrove and Jones, "The Challenge of Mandatory Evacuation: Providing For and Deciding For."

32. Sorensen, "When Shall We Leave? Factors Affecting the Timing of Evacuation Departures"; Bartel Van de Walle and Murray Turoff, "Decision Support for Emergency Situations," *Information Systems and E-Business Management* 6, no. 3 (2008): 295–316.

33. Lori J. Dotson and Joe A. Jones, *Identification and Analysis of Factors Affecting Emergency Evacuations*, Division of Preparedness and Response, Office of Nuclear Security and Incident Response, US Nuclear Regulatory Commission, 2005; Robert M. Stein, Leonardo Dueñas-Osorio, and Devika Subramanian, "Who Evacuates When Hurricanes Approach? The Role of Risk, Information, and Location," *Social Science Quarterly* 91, no. 3 (2010): 816–834.

34. Sims and Baumann, "Educational Programs and Human Response to Natural Hazards"; Baker, "Hurricane Evacuation Behavior"; Lindell, Lu, and Prater, "Household Decision Making and Evacuation in Response to Hurricane Lili"; Stein, Dueñas-Osorio,

and Subramanian, "Who Evacuates When Hurricanes Approach? The Role of Risk, Information, and Location."

35. "National Association of State Departments of Agriculture Emergency Response Plan: Mass Evacuation Incident Annex."

36. Cutter and Smith, "Fleeing from the Hurricane's Wrath: Evacuation and the Two Americas"; M. S. Mannan and D. L. Kilpatrick, "The Pros and Cons of Shelter-in-Place," *Process Safety Progress* 19, no. 4 (2000): 210–218.

37. Cova et al., "Protective Actions in Wildfires: Evacuate or Shelter-in-Place?" *Natural Hazards Review* 10, no. 4 (2009): 151–162.

38. Sorensen, "When Shall We Leave? Factors Affecting the Timing of Evacuation Departures"; Mannan and Kilpatrick, "The Pros and Cons of Shelter-in-Place."

39. Horst Rittel and Melvin Webber, "Dilemmas in a General Theory of Planning," *Policy Sciences* 4, no. 2 (1973): 155–169.

40. It must be emphasized that this summary of factors is largely conceptual; that is, it allows for an appropriate interpretation of which disasters would warrant an evacuation.

41. Fairchild, Colgrove, and Jones, "The Challenge of Mandatory Evacuation."

42. Daniel Henstra, "Evaluating Local Government Emergency Management Programs: What Framework Should Public Managers Adopt?," *Public Administration Review* 70, no. 2 (2010): 236–246.

43. Fairchild, Colgrove, and Jones, "The Challenge of Mandatory Evacuation."

44. This structure was initially put into place by the Disaster Relief Act Amendments of 1974, frequently referred to as the Stafford Act. This act bestowed the national government with power to aid in disaster relief on the conditions that (1) a governor requests such assistance, (2) the degree of necessary response does indeed surpass state capacity, and (3) a governor must put into effect the state's emergency plan. Additionally, federal courts also maintain the authority to order evacuations during severe emergencies "National Association of State Departments of Agriculture Emergency Response Plan: Mass Evacuation Incident Annex," http://www.fema.gov/pdf/emergency/nrf/nrf_massevacuationincidentannex.pdf; Michael McGuire and Debra Schneck, "What if Hurricane Katrina Hit in 2020? The Need for Strategic Management of Disasters," *Public Administration Review* 70, no. s1 (2010): s201–s207; Fairchild, Colgrove, and Jones, "The Challenge of Mandatory Evacuation"; Francis X. McCarthy, *Federal Stafford Act Disaster Assistance: Presidential Declarations, Eligible Activities, and Funding* (Washington, DC: Library of Congress, Congressional Research Service, 2011); *Thames Shipyard and Repair Co. v. United States* (2003), OpenJurist, accessed February 6, 2012, http://openjurist.org/350/f3d/247/thames-shipyard-and-repair-company-v-united-states.

45. Fairchild, Colgrove, and Jones, "The Challenge of Mandatory Evacuation."

46. Connecticut Division of Emergency Management and Homeland Security, *Master Template for Town Updates*. data received March 1, 2013. The Connecticut coastal rise is dramatic. Moving inland from the coast, higher elevations are reached quickly, meaning that the most dangerous places are along the coast. The coastline is attractive, drawing many people to live near it; some communities have seawalls and do not permit construction along the beach.

47. In Bridgeport, approximately eight thousand residents visited shelters (for short or long terms) in the days preceding and following the storm.

48. Approximately 530 people were sheltered in the Town of Greenwich, the highest number ever recorded in a disaster.

49. Shelters contained those from evacuation zones who did not have other places (with friends or family) to go, those from nonevacuation zones who suffered damage to their homes, and people traveling through a locale when the disaster struck. Hotels along the southern Connecticut coast were completely booked through Hurricane Sandy.

50. U.S. Army Corps of Engineers, *Worst Case Hurricane Surge Inundation for Connecticut* (2008).

51. FEMA, "Hurricane Sandy: Timeline"; *Super Storm Sandy After Action Report* (Hartford, CT: Department of Emergency Services and Public Protection: Division of Emergency Management and Homeland Security, January 25, 2013).

52. FEMA, *Super Storm Sandy After Action Report.*

53. FEMA, "New England: One Year after Hurricane Sandy," accessed April 2, 2014, http://www.fema.gov/news-release/2013/10/29/new-england-one-year-after-hurricane-sandy.

54. Connecticut Division of Emergency Management and Homeland Security, *Master Template for Town Updates.*

55. Media reports from various online local news sources, retrieved April 2014; Connecticut Division of Emergency Management and Homeland Security, *Master Template for Town Updates.*

56. Susan L. Cutter, Lindsey Barnes, Melissa Berry, Christopher Burton, Elijah Evans, Eric Tate, and Jennifer Webb, "A Place-Based Model for Understanding Community Resilience to Natural Disasters," *Global Environmental Change* 18, no. 4 (2008): 599.

PART II THE DAYS AFTER THE STORM

3 • OVERLOOKED IMPACTS OF HURRICANE SANDY IN THE CARIBBEAN

ADELLE THOMAS

Hurricane Sandy originated in the Caribbean Sea in October 2012 and hit a number of Caribbean nations hard, including Haiti, Cuba, and the Bahamas. Overall there were nearly eighty deaths in the Caribbean and damages in the hundreds of millions of dollars.[1] However, these impacts in the Caribbean were largely overshadowed in the global media by the impacts of the storm in the northeastern United States.[2] The storm's unusual characteristics (described by Steven Decker and David Robinson in chapter 1) added to this media interest. Even after Hurricane Sandy had passed the Caribbean and the extent of destruction was becoming evident, much media coverage focused on tracking the storm as it made its way toward the United States and involved speculating about potential impacts in the Northeast rather than investigating the toll that the storm had taken on the Caribbean region. The storm's extreme damage to the United States sustained that media interest. Indeed, one British newspaper estimated that less than 20 percent of American and British coverage of Hurricane Sandy focused on the Caribbean, with the vast majority of reporting concentrated on the United States.[3]

This differential emphasis on the effects of Sandy is perhaps unsurprising, given past coverage of extreme events that have hit both the Caribbean and the United States. For instance, Hurricane Andrew, a devastating event in the United States in 1992, also had dire impacts for the Bahamas. The storm caused two fatalities and damages of approximately a quarter of a billion dollars in the Bahamas.[4] As the first major hurricane to affect the nation in decades, Andrew was also deeply unsettling to many Bahamians. However, the majority of global media coverage on Hurricane Andrew focused on the southern United States. This lack of post-storm reporting out of the Caribbean has consequences for people inside and outside the region. As will be discussed, global media coverage of Sandy's effects in the Caribbean provoked mixed reactions in the region that illuminate its diversity and its complex relationships with individuals and organizations from outside. On one hand, Haitian leaders worried that the small number of reports circulating about damage in Haiti would fail to alert potential donors to that country's needs for material aid. On the other hand, officials and people in the tourism industry in the Bahamas sought to reassure visitors that their beaches and hotels were intact soon after the storm.

The discrepancy between media coverage of extreme events that affect both the United States and the Caribbean can be explained in part with the core-periphery model. This model posits that some countries, cities, or regions possess economic, physical, or human advantages that allow them to develop more rapidly than other areas.[5] Core countries are characterized by high employment, wages, and literacy rates, along with effective communication and transportation networks. Conversely, peripheral countries face a number of challenges, including high levels of debt, low life expectancy, and lack of accountable government systems. These characteristics make peripheral areas more physically and socially vulnerable to hazards, given that they have relatively little in resources to anticipate, prepare for, and recover from disasters. Core areas may be less vulnerable to hazards but may face greater economic damages when a disaster does strike, given the relatively large amount of capital and infrastructure that may be affected by a disaster.

This core-periphery model also applies to media attention. Studies have found that the level of development of a country correlates with the amount of media coverage given post-disaster.[6] Developing countries, that is, those in the periphery, are afforded less media coverage compared with developed or core countries. When the magnitude of the disaster is accounted for, global media coverage gives more time and attention to disasters occurring in more developed countries over less developed countries.[7] Put more bluntly, McLurg's Law, named after a British news editor, uses a scale to quantify the newsworthiness of a disaster.[8] For Britain, the scale means that one dead Briton is worth five dead Frenchmen, twenty dead Egyptians, five hundred dead Indians, and one thousand dead Chinese. While this is a macabre analysis, it portrays the relationship between where disasters occur and the coverage that they are afforded. Countries in core regions have significantly higher chances of being in the news than countries in peripheral regions.[9]

Media coverage of disasters in peripheral areas may seem of interest only to the residents of those areas, but underreporting from these areas also affects whether we see sea level rise and new patterns in hurricane formation as problems that the entire world is facing and must solve together. In addition, these peripheral areas contribute a far lower percentage of greenhouse gas emissions that drive climate change, in comparison with core areas.[10] However, many of these peripheral areas are disproportionately affected by the impacts of climate change, including sea level rise and increased intensity of extreme events.

This chapter aims to shed light on some of the impacts and vulnerabilities in the Caribbean countries affected by Sandy that were given minimal media attention elsewhere. Impacts within the archipelago nation of the Bahamas are also analyzed to provide a case study of how a focus on economic development and reliance on specific economic sectors, specifically tourism, affects vulnerability and responses to extreme events.

COMPOUND IMPACTS AND NEW VULNERABILITIES: HAITI AND CUBA

Although most island nations in the Caribbean can be considered developing countries, there is a considerable range of wealth and institutional development across these countries that make some of them more vulnerable to extreme events than others. Most of the deaths and damage from Sandy occurred in two countries with very different characteristics, Haiti and Cuba. Haiti, a democratic nation, is considered to be one of the poorest countries in the Western Hemisphere, with a gross domestic product (GDP) per capita of US$1,300 and a population of about 10 million people.[11] On the other hand, Cuba has a similar population of about 11 million people, but has a socialist government and a higher GDP per capita of US$10,200.[12]

The differences in governmental structure, levels of poverty, past experiences with extreme events, and approaches to preparing for the storm had implications for the damages and deaths in each country.

Haiti, the Caribbean nation most badly harmed by Hurricane Sandy, provides an example of the effect of multiple extreme events in a short time span. Sandy was the latest in a series of extreme events that have damaged Haiti recently. In 2010, a devastating magnitude 7.0 earthquake (Richter scale) shook the island nation, resulting in more than three hundred thousand deaths, damaging the homes and livelihoods of more than 3 million people and destroying much of the infrastructure of the capital city.[13] This catastrophic earthquake overwhelmed Haiti, exposing the tremendous vulnerability of its population to extreme events. International aid was a major component of recovery as the country relied on donations and volunteers from around the world to aid in repairing the destruction, with the U.S. government alone providing over US$1.1 billion in aid. In 2012, just two months prior to Hurricane Sandy, Tropical Storm Isaac hit Haiti, resulting in twenty-four deaths, damage to infrastructure, and more than US$24 million in agricultural losses.[14] The combination of both the tropical storm and the earthquake resulted in approximately four hundred thousand Haitians being housed in refugee camps in mid-2012. A cholera epidemic had afflicted half a million people since the earthquake and was exacerbated by the tropical storm, resulting in an increase in the number of people affected by the infection.

When Hurricane Sandy hit Haiti in late 2012, it therefore affected a nation that was already weakened by these prior extreme events. During the passage of Hurricane Sandy, the majority of refugee camps experienced flooding or leaking of sewage systems, which led to unsanitary conditions and once again worsened the cholera epidemic. Sandy is estimated to have added at least two hundred thousand more homeless people to the existing four hundred thousand who were already housed in temporary refugee camps.[15] The storm destroyed more than 70 percent of the nation's crops and killed many livestock, putting more than 1.5 million Haitians at risk of malnutrition. Approximately sixty deaths were also attributed to the storm. These impacts, during a time of recovery from prior extreme events, increased the likelihood of Haiti's falling deeper into chaos and civil unrest.[16]

The passage of Hurricane Sandy highlighted existing vulnerabilities in Haiti. The storm exposed the compound effects of a series of extreme events. The inability to rapidly recover from the impacts of a disaster makes the country more vulnerable to subsequent extreme events. It then becomes ever more difficult and expensive to recuperate. For example, the exacerbation of the cholera epidemic mostly affected those who were already located in refugee camps.

Hurricane Sandy also exposed emerging vulnerabilities within the Caribbean region, most notably in Cuba, where eleven people died and more than two hundred thousand homes were damaged.[17] This was surprising, given Cuba's reputation for excellent hurricane preparation and mass evacuations. Cuba's approach to hurricane preparation and recovery has made it one of the Caribbean nations least harmed by prior hurricanes and proof that institutional preparation and social awareness may be more important than wealth in determining how a country can reduce hazards.[18] However, Sandy exposed vulnerabilities in Cuba's hurricane management approach. Previous storms have mostly affected the western portion of the island, and these areas were prepared for the storm. Sandy caught some eastern Cuban communities unprepared, however. Santiago de Cuba, located on the eastern side of the island, which was severely damaged by the storm, had not experienced any hurricane impacts in the past sixty years.[19] The coastal town of Aguadores, also on the eastern side of the island, was almost completely demolished by the storm. Hurricane Sandy also caused damage to crops, thereby affecting the food security of the nation, though this damage is less related to preparedness. Overall, approximately US$80 million in damages was attributed to Hurricane Sandy in Cuba.

In Cuba, Hurricane Sandy exposed the need for the nation to raise community awareness and preparedness for disasters to all residents of the island, not just those in a particular geographic location. The storm also highlighted how past experiences with extreme events affects future preparedness and awareness. Settlements that had limited impacts from storms in recent times were ill prepared for Hurricane Sandy.

The lack of global media attention to these two countries had negative consequences. The media attention paid to Haiti was far less than after the 2010 earthquake, although there were dire impacts from Sandy in this struggling country. This lack of coverage led the Haitian prime minister to appeal for greater attention to the troubles of the nation and to call for support to save lives and property after Hurricane Sandy's passage.[20] A lack of international attention limited the aid that the country received, thereby extending the time needed to recover and increasing vulnerability to future extreme events. In Cuba, the lack of coverage of deaths and damage to the country's eastern side meant an opportunity was missed for others to learn about the limits of hazard reduction and disaster response systems that emerge as adaptations to prior storms and hazards.

BACK TO BUSINESS: THE BAHAMAS

The lack of media coverage of the Bahamas had yet another set of consequences. While both Cuba and Haiti have high populations and medium or low GDP per capita, respectively, the Bahamas is one of the highest-income nations in the region, with a GDP per capita of US$32,000, and a small population of about 370,000.[21] Yet Hurricane Sandy also had dire impacts in the Bahamas, and as this case study shows, vulnerability to extreme events within the region is not solely determined by the level of economic development or by dense populations. The Bahamas, a democratic nation, illustrates how economic development and reliance on specific economic sectors affects vulnerability and responses to extreme events.

The Bahamas is an archipelago nation comprising a chain of more than seven hundred islands and cays in the Atlantic Ocean. The capital of the country, Nassau, is located on New Providence, a small island near the center of the island chain. New Providence is home to approximately 60 percent of the nation's population and is the center of government and the majority of economic activities.[22] The other islands of the nation are collectively referred to as Family Islands and have much lower population densities and much less economic activity. The Bahamas is highly dependent on tourism, which provides more than 50 percent of jobs and GDP for the country. The majority of tourism income is also concentrated in the capital, with the bulk of air and cruise tourists visiting New Providence. This reliance on tourism and the concentration of the industry in New Providence greatly influenced the response of the nation to the effects of Hurricane Sandy.

In the Bahamas, Hurricane Sandy exposed new vulnerabilities, as it did in Cuba. New Providence suffered damage in areas essential to the tourism industry. Key infrastructure along the coastline was affected, including the international airport, the cruise ship harbor, and key road transportation routes and facilities.[23] Flooding inundated coastal areas, caused damage to hundreds of homes and millions of dollars of damage to key hotel infrastructure, including the famous Atlantis resort. These impacts were surprising, as the island had largely escaped unscathed from previous hurricanes. Indeed, a common perception among Bahamians was that the capital island of New Providence was not at risk of hurricane damage because of its geographical location at the center of the archipelago, located between other islands.[24] The Family Islands were thought to be more vulnerable to hurricanes because they are not geographically shielded by other islands and have in the past been more affected by hurricanes. However, Sandy exposed the vulnerabilities of New Providence as a small, flat island with dense coastal development.

On the Family Islands, a number of coastal communities were certainly also damaged by coastal flooding, with storm surges reaching more than twenty-five feet in some areas.[25] Difficulties in accessing services there indicate the peripheral nature of the Family Islands. On some islands, flooding prevented vehicular passage to many outlying communities.[26] This prevented access to the limited health-care services available on the Family Islands, while the storm's atmospheric conditions also barred any emergency air travel to better health care in New Providence. Following the passage of Hurricane Sandy in the Bahamas, there were two deaths, almost one thousand homes that were deemed uninhabitable, and more than US$700 million in damage, about 9 percent of the GDP of the country.[27] These impacts were viewed by the public as surprising, given that Hurricane Sandy was only a Category 1 storm as it passed the Bahamas, and led to speculation about the damage that might have occurred if Sandy had been a Category 4 or 5.

However, despite the serious implications of Sandy for the country, the Bahamian government focused on sending out the message to the international community that the Bahamas was minimally affected by the storm and open for business. There was a particular urgency to deliver this message before the American public focused on how Sandy would affect the United States. "As of Friday afternoon, the American consumers are focusing on Sandy entering to the northeast, which means that it is travelling away from the Bahamas and the Caribbean. But before that happens we want to ensure that the message that we are back open and ready for business is with them," stated the director of the national emergency agency for the country.[28] While Haiti was sending out a plea for international help and trying to publicize its plight, the Bahamas purposely sent out the message that the country had rapidly recovered from any impacts of the storm and was ready for tourists.

The dependence on tourism motivated efforts to quickly recover from the storm. In New Providence, damage to infrastructure at the airport and docks was quickly repaired to facilitate the continued influx of air and cruise ship tourists.[29] A major thoroughfare that connects the airport with the majority of hotels and tourist attractions experienced serious damage, including the collapse of a section of the road as well as inundation with sand, rocks, and storm surge, making the road impassable at a number of points. The focus of recovery was to repair this road as soon as possible to ensure that traffic and tourism could function as usual.

However, the quick repair and return to pre-Sandy conditions failed to acknowledge the vulnerabilities that Sandy exposed in New Providence. The major thoroughfare that was damaged by Sandy had been similarly damaged by other tropical storms, although this collapse was the most serious in recent times.[30] Returning to pre-Sandy conditions without addressing the need to possibly redesign, reroute, or protect the road in some other way did not take into account the vulnerabilities that were being exposed. Continued use of the thoroughfare without significant changes to reduce its vulnerability will likely lead to ongoing and perhaps more serious impacts in the future, given rising sea levels and the potential for more intense, extreme storm effects. This vulnerability has implications for not only the tourism industry but also for residents of the area, especially as population increases.

While it was back to business in New Providence, a different approach to recovery was taken in the other islands. In the Family Islands, the vulnerabilities that Sandy exposed were acknowledged, and there was a focus on the need to improve preparation and response to extreme events. A number of policy proposals such as mandatory evacuations of remote communities, aviation plans for emergency airlifts, mandatory elevation of infrastructure along the coast and even the need to abandon some coastal communities have been discussed in the months since Sandy.[31] Structural assessment of major thoroughfares on the Family Islands was conducted to assess any redesigning needs. Coastal communities that had been previously damaged by tropical storms began to review the feasibility of relocating further inland to prevent continued destruction from future extreme events.[32] The impacts of climate change were also included in discussions as residents and policymakers reviewed the possibility of increased intensity of hurricanes and estimated how this would heighten the vulnerability of communities in the future.

These responses and dialogues that were taking place in and about the Family Islands showed that the impacts of Hurricane Sandy on the Bahamas were far from minimal and that there were a number of vulnerabilities exposed. However, discussion of the need to decrease vulnerability and learn from the storm did not take place on the same scale in New Providence. While New Providence is the most densely populated and economically significant island in the archipelago, the need to recognize vulnerabilities and adapt accordingly was unfortunately displaced by the need to quickly return to normal business operation. This difference may be due to the sparse population on the Family Islands, which may make it less complex to implement significant changes such as relocation of communities. However, the lack of attention to adaptation in New Providence places a large proportion of the nation's population at increased risk to extreme events.

It was beneficial for the nation that international tourists view New Providence as unaffected and open for business, thus overshadowing the need for adaptation. While this approach may be profitable for the tourism industry and short-term economic gains, it is problematic when considering vulnerability on a longer time scale, given the number of people and amount of economic activity at risk as a result of the lack of adaptation.

CONCLUSION

Hurricane Sandy was a devastating storm that wrought havoc across the Caribbean and the United States. Although the majority of media coverage has focused on implications of Sandy in the United States, it is important to recognize the impacts that the storm had for the Caribbean region. Hurricane Sandy highlighted existing vulnerabilities in Haiti, showing the difficulties that the nation faces as it attempts to recover from a series of extreme events with limited resources and reliance on external aid. The storm exposed new vulnerabilities in Cuba, showing the need for more geographically comprehensive approaches to vulnerability assessment and adaptation. In both nations, Sandy highlighted how governmental structure, economic development, and past experience with storms shape vulnerability to storms, influence deaths and damage from extreme events, and also compromise the food security of millions in the region.

For the Family Islands in the Bahamas, Hurricane Sandy appears to have been an eye-opener to the need to reduce disaster vulnerability. However, for New Providence, attention to these issues of vulnerability and adaptation has unfortunately not been considered in depth, given the focus on maintaining a lucrative and uninterrupted tourism business. The response to the storm in New Providence is strikingly similar to the responses in the New Jersey and New York shore areas (as discussed by Briavel Holcomb in chapter 9) and in lower Manhattan, where we observe a failure to reflect and adapt as a consequence of rebuilding quickly and broadcasting the message that these places were back to business. The disconnect between responses in New Providence and the Family Islands is most unfortunate, given that New Providence is home to more than half of the Bahamian population and the center of tourism and government. Although there were initial hopes by some residents that Sandy would be a transformative event to decrease the long-overlooked vulnerability of New Providence to extreme events, there was a rush to return to pre-Sandy conditions and a failure to address exposed vulnerabilities.

The case study of the Bahamas also shows that detailed consideration of economic forcing factors is essential when assessing the potential for reduction of future vulnerability. It is difficult to acknowledge vulnerabilities and make plans to reduce risk when the main objective is to restart economic activities as soon as possible. There must be a balance between disaster risk reduction and economic stability, perhaps by framing restoration in terms of phases, with immediate restoration of services in a manner that facilitates adaptation over a longer period. These are important and interconnected issues, because reducing hazards ultimately improves economic stability by making essential industries, such as tourism, less vulnerable to future extreme events.

LESSONS

Several lessons emerge from this analysis of varied impacts and responses in the Caribbean:

- Slow recovery from impacts of extreme events greatly increases vulnerability to future extreme events.

This is evidenced by the case of Haiti, which had not fully recovered from the 2010 earthquake and 2012 tropical storm. More than four hundred thousand people were in refugee camps when Hurricane Sandy hit the nation.

- Community awareness and preparedness and prior experience with extreme events reduces vulnerability.

Although Cuba is known for hurricane preparation, residents in eastern Cuba had little experience with prior hurricanes and were thus ill prepared when Sandy hit.

- Media attention after extreme events affects perceptions of damage and international donations.

Lack of media attention for Haiti led for appeals from the nation for greater international support. However, lack of media attention for the Bahamas bolstered the perception that the destination was open for tourism business.

- Reliance on a narrow set of industries affects responses to extreme events.

The Bahamas, highly dependent on international tourism, focused on quickly responding to impacts of Hurricane Sandy in areas that were essential to the industry in order to facilitate uninterrupted tourism services. A more measured analysis and response may have made for a more resilient Bahamian nation, one better able to adapt to the pressures of both a growing population and climate change.

- Recovery from extreme events should incorporate actions that decrease vulnerability to future events rather than simply returning to pre-event conditions.

Returning to pre-event conditions ignores exposed vulnerabilities and may actually increase vulnerability to future events. For example, continued coastal development in the Bahamas after Hurricane Sandy may expose the nation to greater impacts from future extreme events and from sea level rise.

NOTES

1. United Nations Development Programme, *Hurricane Sandy Kills around 80 in the Caribbean, 1.8 Million Affected in Haiti* (New York, 2012), http://www.undp.org/content/undp/en/home/presscenter/articles/2012/11/02/hurricane-sandy-kills-around-80-in-the-caribbean-1–8-million-affected-in-haiti.html.
2. Garry Pierre-Pierre, "Hurricane Sandy: It Hit the Caribbean Too, You Know," *Guardian*, November 2, 2012.
3. Roy Greenslade, "Hurricane Sandy: Why Have the British Media Ignored Caribbean Victims?," *Guardian*, October 30, 2012.
4. United Nations Department of Humanitarian Affairs, *Hurricane Andrew Aug 1992 UN DHA Information Reports 1–3*, August 26, 1992, http://reliefweb.int/report/bahamas/bahamas-and-usa-hurricane-andrew-aug-1992-un-dha-information-reports-1–3.
5. Paul Krugman and Anthony Venables, "Globalization and the Inequality of Nations," *Quarterly Journal of Economics* 110 (1995): 857–880.
6. Eleanor Singer, Phyllis Endreny, and Marc Glassman, "Media Coverage of Disasters: Effect of Geographic Location," *Journalism and Mass Communication Quarterly* 91 (1991): 48–58.
7. Douglas Belle, "*New York Times* and Network TV News Coverage of Foreign Disasters: The Significance of the Insignificant Variables," *Journalism and Mass Communication Quarterly* 77 (2000): 50–70.
8. Denis McQuail, *Mass Communication Theory: An Introduction* (London: Sage, 1987).
9. Tsa-Kuo Chang, "All Countries Not Created Equal to Be News: World System and International Communication," *Communication Research* 25 (1998): 528–563.
10. Intergovernmental Panel on Climate Change, "Summary for Policymakers," in *Climate Change 2007: Impacts, Adaptation, and Vulnerability; Contribution of Working Group II to the Fourth Assessment Report of the Intergovernmental Panel on Climate Change*, ed. M. L. Parry, O. F. Canziani, J. P. Palutikof, P. J. van der Linden and C. E. Hanson (Cambridge: Cambridge University Press, 2007), 7–22.
11. "CIA World Factbook: Haiti," https://www.cia.gov/library/publications/the-world-factbook/geos/ha.html.
12. "CIA World Factbook: Cuba," https://www.cia.gov/library/publications/the-world-factbook/geos/cu.html.
13. U.S. Agency for International Development (USAID), "Haiti—Earthquake and Cholera: Fact Sheet #12" (2011), http://www.usaid.gov/crisis/haiti/haiti-disaster-response-archive.
14. USAID, "Haiti—Hurricane Sandy: Fact Sheet #1" (2013), http://www.usaid.gov/sites/default/files/documents/1866/02.15.13%20-%20Haiti%20Hurricane%20Sandy%20Fact%20Sheet.pdf.
15. Ibid.

16. Jonathan Watts, "Caribbean Nations Count Cost of Hurricane Sandy," *Guardian*, October 29, 2012.
17. Sarah Rainsford, "Hurricane Sandy: Cuba Struggles to Help Those Hit," *BBC*, November 12, 2012.
18. Pedro Mas Bermejo, "Preparation and Response in Case of Natural Disasters: Cuban Programs and Experience," *Journal of Public Health Policy* 27 (2006): 13–21.
19. International Federation of the Red Cross, "Cuba: The Struggle to Recover from Hurricane Sandy Seven Months On," November 5, 2013, http://www.ifrc.org/en/news-and-media/news-stories/americas/cuba/cuba-the-struggle-to-recover-from-hurricane-sandy-seven-months-on-63669.
20. United Nations Stabilization Mission in Haiti, "Press Release—Haiti: Government and Humanitarian Agencies Call for Additional $39M in Aftermath of Hurricane Sandy," November 8, 2012, http://www.minustah.org/press-release-haiti-government-and-humanitarian-agencies-call-for-additional-39m-in-aftermath-of-hurricane-sandy.
21. "CIA World Factbook: The Bahamas," https://www.cia.gov/library/publications/the-world-factbook/geos/bf.html.
22. Adelle Thomas, "An Integrated View: Multiple Stressors and Small Tourism Enterprises in the Bahamas" (PhD diss., Rutgers University, 2012).
23. Neil Hartnell, "IDB: Government Estimates Sandy Costs at $703M," *Tribune* [Nassau], January 21, 2013.
24. Thomas, "An Integrated View."
25. Lindsay Thompson, "Prime Minister Accompanies NEMA to Hurricane Sandy Impacted Areas," *Bahamas Weekly*, October 29, 2012.
26. Denise Maycock, "Calls for Repairs to Fishing Hole Road," *Tribune*, November 12, 2012.
27. Ava Turnquest, "Hurricane Sandy Damage Adds Up to More Than $700M," *Tribune*, January 21, 2013.
28. Dana Smith, "The Bahamas Back in Business after Hurricane Sandy," *Tribune*, October 27, 2012.
29. Denise Maycock, "Grand Bahama Airport Reopens," *Tribune*, October 30, 2012.
30. Neil Sealey, "West Bay Street, Nassau, 2012: Sand Loss from Beaches and Damage to Infrastructure," http://www.academia.edu/7066871/Hurricane_Sandy_damage_in_Nassau_Bahamas_2012.
31. Communication by the Rt. Hon. Prime Minister and Member for Centerville Regarding Assessment Report on Hurricane Sandy, November 1, 2012, http://www.bahamaislandsinfo.com/index.php?option=com_content&view=article&id=13037:pm-gives-communication-in-parliament-on-aftermath-of-hurricane-sandy&catid=34:Bahamas%20National%20News&Itemid=147.
32. "Storm Floods Frightening," *Tribune*, October 29, 2012.

4 • POLLING POST–HURRICANE SANDY

The Transformative Personal and Political Impact of the Hurricane in New Jersey

ASHLEY KONING AND DAVID P. REDLAWSK

On October 29, 2012, Hurricane Sandy made landfall near Brigantine, New Jersey, wreaking havoc on the New Jersey and New York coastal areas, as well as much of the inland, before it turned north to head through New York and New England. It caused eighty-seven deaths and an estimated $78.8 billion worth of damage in New Jersey and New York alone.[1] The hurricane also affected the way in which many residents now view their own future and the repercussions of future events like Sandy.

The storm not only made dramatic changes in the physical landscape (one town on a barrier island was literally cut in half by surging waters); it transformed the political landscape as well. Sandy may have changed the course of the final days of the 2012 presidential race, seemingly solidifying a victory for President Barack Obama by giving him the chance to exercise his presidential powers in the storm's aftermath. Sandy also brought Governor Chris Christie into the national political spotlight as a symbol of leadership and bipartisan cooperation, amassing him immense power and popularity that continued for more than a year after the storm, a rare feat for any politician, particularly in the wake of a natural disaster.

This chapter explores the personal and political consequences of Sandy through surveys conducted since the storm's immediate aftermath by the Rutgers-Eagleton Poll, the New Jersey statewide public opinion poll of the Eagleton Center for Public Interest Polling at Rutgers University.[2] The extensive polling reported below was not matched in other affected states, so it stands out as an important record of reactions in one of the states most shaken by this major storm. The first section details personal responses to the storm, such as how it impacted New Jersey residents' emotions, behaviors, attitudes, and overall lives. The second section highlights the political consequences of Sandy, particularly its influence on the 2012 presidential election and on Christie's political legacy.

While treated separately here, these two stories about the personal and the political effects of the hurricane are intertwined. The direct impact, personal perceptions, future hopes, and emotions expressed by individuals about Sandy influenced how local, state, and federal level leaders responded to the storm and, in turn, how citizens responded to and chose their leaders in a post-Sandy world. Our findings show that with attention galvanized around the destruction along the state's cherished shoreline, many New Jerseyans sought comfort from Governor Christie soon after the storm, raising him to unprecedented political heights.

PERSONAL RAMIFICATIONS OF SANDY'S INITIAL IMPACT AND DAMAGE

In a survey taken within three weeks of Hurricane Sandy's landfall, the Rutgers-Eagleton Poll found that two-thirds of New Jersey residents were personally affected by the storm.[3] Nearly every resident who reported being at all affected by the storm lost power for at least some time in its aftermath; 65 percent of those affected were without electricity for more than five days. One-fifth of affected residents were forced to leave their homes, and almost one-third in affected areas reported damage to their houses or other property.

The impact and immediate aftermath of Sandy was especially serious for exurban and shore residents, the regions hardest hit with dangerous winds and flooding.[4] As figure 4.1 indicates, nearly a quarter of all residents of shore counties reported being forced to leave their homes, and those residents were far less likely than others to be able to return to their homes in the first three weeks after the storm. Residents of shore counties and the hilly, tree-covered northwest part of the state were most likely to report damage to their houses or other property, owing to the storm's track.

New Jersey has a history of big hurricanes, but because they occur irregularly, Sandy took New Jerseyans by surprise. In the first few weeks after Sandy, only 35 percent of New Jersey residents thought the state was adequately prepared. The storm changed behavioral intentions, though whether it changed actual behavior is less clear from surveys. A majority of residents in March 2013 said they were now more likely to formulate evacuation plans, prepare an emergency kit or bag, and purchase flashlights and candles for future events like Sandy. Almost half of respondents statewide said they would be more likely to purchase a generator. These precautionary measures were embraced most by Sandy's hardest-hit regions.[5]

Aftermath and Recovery

In New Jersey, where there is usually no shortage of cynicism about government, residents were initially very positive about the response by political officials, first responders, electric companies, and assistance agencies and organizations like the Federal Emergency Management Administration (FEMA) and the Red Cross, with a majority saying they had all done well with handling the crisis and recovery. Large majorities thought state and local government acted efficiently and responded in the right amount of time, though they did not express the same sentiment for the federal government, where congressional Republicans delayed relief funding to the region.[6]

New Jerseyans reported help and recovery efforts from a combination of sources. In a poll three months after the storm, only 13 percent of those affected claimed to have used any type of disaster assistance from organizations like FEMA or the Red Cross.[7] New Jerseyans had greater success with their own insurance companies: in the first few months after Sandy, a third of residents claimed they had reported damages to their insurance company, and almost three-quarters of those who did had already received payment for damages by that time.

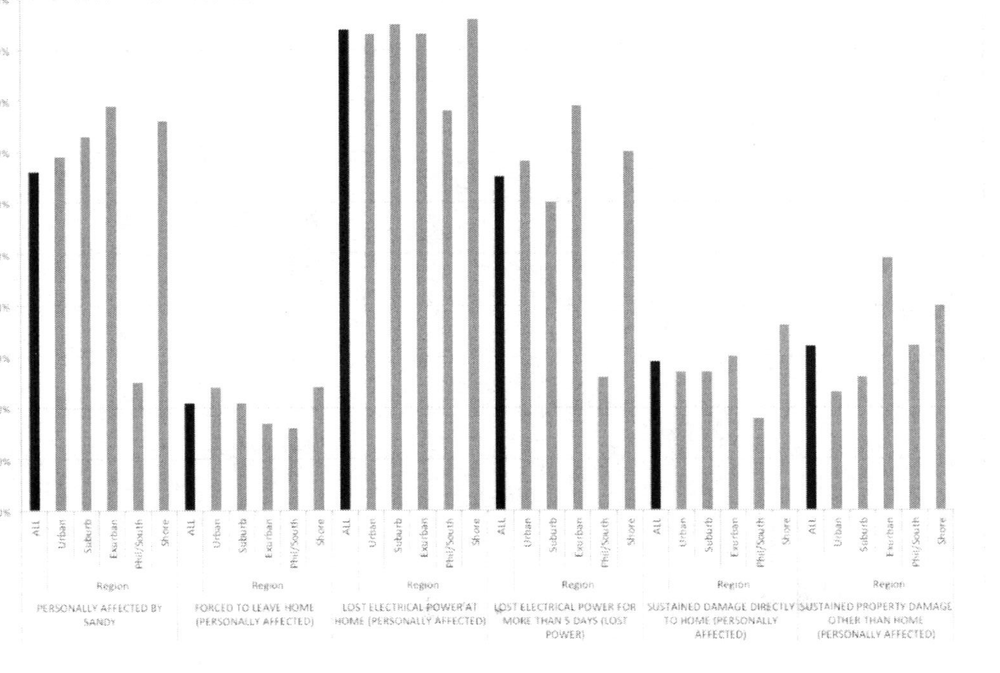

FIGURE 4.1. Questions asked of New Jersey adults. "Superstorm Sandy recently did severe damage across the Northeast and especially in New Jersey. Were you, yourself, personally affected by the storm?"; "Were you forced to leave your home?"; "Did you lose your electrical power at home?"; "Did you lose your electrical power for more than five days?"; "Did you sustain damage directly to your home?"; "Did you have property damage other than to your home?" (Source: Rutgers-Eagleton Poll, Eagleton Center for Public Interest Polling, November 2012.)

Those most affected by the storm grew more discouraged and frustrated with Sandy aid and assistance one year out. In a Monmouth University Polling Institute panel study of the hardest hit Sandy victims, almost half of those interviewed said they were denied needed services and assistance from New Jersey's federally funded Sandy assistance program, "reNew Jersey Stronger."[8] More than half of these victims also said obtaining information about recovery or rebuilding from the state government and the federal government had been somewhat or very difficult, and similar numbers had not found either to be helpful in their recovery process in general. Three-quarters felt they had been "largely forgotten" in the state's recovery process.

Therefore, despite their initial accolades, New Jerseyans did not respond as positively to recovery efforts as time went on, despite the state government's promotion that the state was back in business and, as the now famous and all-too-catchy jingle stated, "Stronger than the storm."[9] Just before the first post-Sandy Memorial Day weekend, Governor Christie graded "conditions on the boardwalk" at the Jersey shore on a scale from 1 to 10, telling Matt Lauer on the *Today Show*, "It's an 8 in some places and a 4 in others so it depends. . . . Here on the boardwalk, people will see an 8 out of 10 when they start coming here in June."[10] By comparison, when asked to give their own rating, residents on average rated shore recovery at 6.4 and state recovery at 7.0.[11] Just after the storm's one-year anniversary, New Jerseyans gave even lower ratings, scoring the shore's recovery at 4.7 and the state's overall recovery at 6.1.[12] They scored recovery for homeowners with sustained damage at 4.8. As we will see later, however, the frustration evident in these scores did not seem to influence how residents viewed Governor Christie.

Rebuilding

The shore looms large in New Jersey culture, and many felt its destruction viscerally. Yet, by February 2013, 62 percent of state residents were cautious about rebuilding at the shore, believing that assessments of the potential for future damage should be made before doing so. This was in stark contrast to the public push by political and business leaders to move forward with recovery efforts as quickly as possible in order to promote summer tourism. Residents also favored implementing strong precautionary measures to combat future natural disasters, particularly in flood zones.[13] The most popular measures were building sand dunes or seawalls and elevating buildings on pilings to lessen future flood damage, both strongly favored by nearly two-thirds of residents. More than half strongly favored moving development further from the waterfront. Residents were less supportive of converting developed land to open space or completely abandoning damaged shore areas. As rebuilding efforts took shape in the first months after Sandy, New Jerseyans overwhelmingly believed that government should pay for all or part of any property damage caused by the hurricane, with only 17 percent saying individuals should have to pay for all property damage costs themselves.

Attitudes toward Climate Change

Even before Hurricane Sandy, New Jerseyans showed concern about global climate change, although that concern ranked lower than the usual concerns about taxes and the economy. An unpublished telephone survey of New Jersey residents by the Rutgers-Eagleton Poll in March 2010 found that 55 percent thought it was very important to "stop global warming" and that about half said, while the financial cost of solving the problem would be high, it would be worth it. Although scientists project that climate change is making it more likely that a storm will become severe, they do not speak of climate change as causing any particular storm. Even so, research by the Yale Project on Climate Change Communication shows that people are much more likely to express belief in global warming when they have personally witnessed extreme weather that differs from their usual weather expectations.[14]

We likewise found that Sandy reawakened concern. Five months after Sandy, nearly two-thirds of New Jerseyans saw global climate change as the likely culprit of the hurricane and of storms in 2011; less than one-third saw the storms as isolated weather events.[15] Just about half were pessimistic about the future, believing it was at least somewhat likely that global climate change would cause another natural disaster like Sandy in their own community in the coming year. Three-quarters said it was at least somewhat likely that climate change would cause a natural disaster somewhere in the United States during that time. Concern about global climate change was still evident more than a year after Sandy, as extreme weather continued to pummel the state.[16] Officials, experts, and others responded to this sentiment with proposals and projects that not only would repair damage from Sandy but also prepare for what may become a more frequent occurrence in the future.[17]

Life after Sandy

Cold, hard statistics such as reported here tend to both enlighten and obscure. We can see how those affected by a storm like Sandy respond, but it is harder to assess the emotional toll such an event can take. One way is to simply ask people whether things are back to "normal."

Three months after the hurricane hit, three-quarters of New Jerseyans thought that life was not yet back to normal. This number has fluctuated little over the time it has been asked, and the time horizon in which people think normality will return did not change in the year following the storm. A majority has consistently thought that a return to pre-Sandy conditions is still one to five years away, even as the time since Sandy has increased.

New Jerseyans overwhelmingly call Sandy a "transformative" event. When asked four months after the storm why Sandy was transformative, many mentioned the unexpectedness and rarity of the event, the magnitude and widespread impact of the storm, and the damage done to the state's beloved shore region in particular. New Jerseyans saw Sandy as a "wake-up call" that "opened people's eyes" and made them more "alert" to the possibility that these types of natural disasters can happen at home as well as in other parts of the country. Residents also said that Sandy "devastated" their home state, calling it an "epic," "rare," and "powerful" event. They mentioned that the hurricane showed clear evidence of climate change, making it a "fact," and that there would be many more storms like this to come. Residents called the event "unanticipated," "unexpected," a "nightmare," and a "lesson" in how to prepare for the future. One individual even said the world in which he grew up did not "exist anymore" post-Sandy and that life would never be the same.

The only positive comments that New Jerseyans had about Sandy was that the hurricane brought residents "closer together" and that the outpouring of help from fellow New Jerseyans and others was a "touching" experience that made residents feel closer to their communities. A word cloud in figure 4.2 illustrates these concerns visually, with the size of each word indicating how often it appeared in answers to this open-ended question.

The emotional effects persist. A year and a half after Sandy, Monmouth University researchers interviewed people who suffered major damage to their primary home.[18] Respondents reported poor mental health, especially among those victims who had not yet been able to return to their homes, those who were economically disadvantaged, and those with unstable

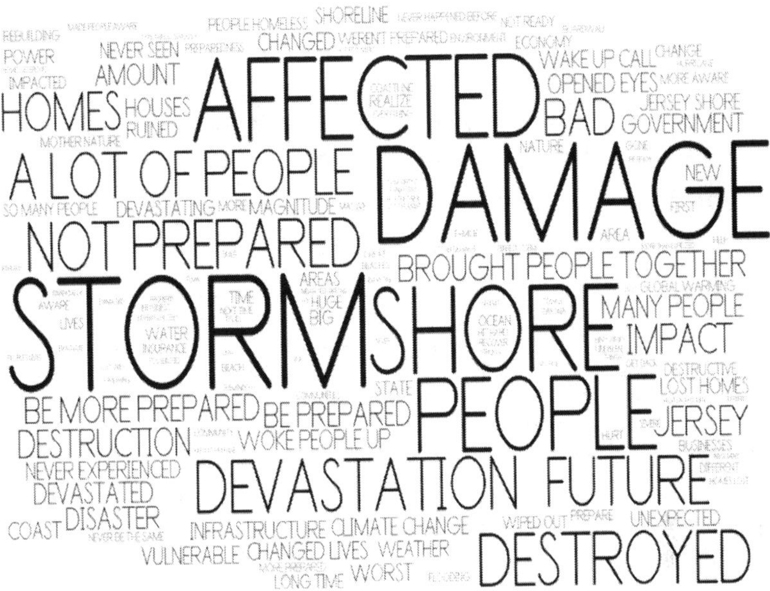

FIGURE 4.2. Question asked of New Jersey adults. “In just a couple words, please tell me why you think Sandy was a transformative event.” Verbatim responses were categorized and grouped with other similar responses. Categories are reflected in the word cloud; word size in the word cloud represents the frequency with which respondents mentioned something pertaining to that category. (Source: Rutgers-Eagleton Poll, Eagleton Center for Public Interest Polling, March 2013.)

family employment. About half of these victims continued to suffer from at least mild psychological stress, much higher than the percentage reported by the state population as a whole.

Overall, the various polls indicate that some people continued to be materially and emotionally affected by the storm many months after it passed, while concern about the state’s vulnerability to storms and climate change was shared by a majority of the broader population.

POLITICAL IMPACT

The storm also brought together President Obama and Governor Christie, two of the most polarizing figures at that time, in a rare moment of bipartisanship, creating a prolonged "rally" effect for Governor Christie and catapulting him, for some time thereafter, into the early status of Republican frontrunner for the 2016 presidential nomination. While Christie's post-Sandy ratings highs have completely vanished in the wake of numerous allegations against his administration as we write this chapter, it seemed for a long time as if Sandy's political changes might be as permanent as its physical ones.

The 2012 Presidential Election

Hurricane Sandy was the "October surprise" of the 2012 presidential election, hitting just a week before Election Day. While immediate safety was at the forefront of everyone's minds, Sandy also sent preelection pollsters, survey researchers, and the media scurrying for at least metaphorical cover. Worries included whether people would be willing to talk to pollsters and whether refusals and the inability to contact others would result in samples that were biased, potentially leading to incorrect observations in tracking polls or long-studied political trends. Some pollsters, moreover, believed that polling would be wholly inappropriate for people caught in the midst of or recovering from a natural disaster.[19] Some polls kept going right up until the storm hit, some encountered challenges as they tried to forge through, and others, particularly statewide polls in those areas most affected, canceled further preelection polling altogether.[20]

President Barack Obama and Republican presidential candidate and former Massachusetts governor Mitt Romney both paused their campaigns.[21] Obama was continually briefed on the storm, visited FEMA headquarters and storm-ravaged territories (his bipartisan boardwalk tour alongside Christie perhaps the most memorable), and assured states of expeditious delivery of aid.[22] Romney reorganized a campaign stop into a donation drive for Sandy victims, but Obama had the advantage of remaining in the political spotlight by assuming his presidential duties. Despite Christie's enthusiastic campaigning for Romney prior to the storm, Christie declined an invitation from the Republican nominee to speak at a nearby campaign event days after Sandy hit. When asked if Romney would visit, Christie famously responded: "I have no idea, nor am I the least bit concerned or interested. I've got a job to do here in New Jersey that's much bigger than presidential politics, and I could [not] care less about any of that stuff."[23]

Sandy's ultimate impact on the 2012 presidential election is debatable, but a comparison of tracking polls, averages, and election forecast models right before and after the storm hint at a "Sandy bump" for Obama.[24] Many believe, including Romney himself, as he would later admit, that Obama's final campaign days spent on top of the polls were, at least in part, a product of his presidential presence in Sandy's aftermath.[25] Yamil Velez and David Martin argue for the importance of Sandy on the voting behavior, and electoral votes, of key swing states, finding that Obama would have lost Virginia to Romney had the storm not hit the state at all and that Obama would have won North Carolina had that state been hit.[26] Obama had always been projected to win New Jersey, but he won by a record 17 points, much greater than predicted pre-Sandy and an increase from his margin in the state in 2008.[27]

Turnout in New Jersey was the lowest in the state's history, with only about 60 percent of registered voters voting.[28] Even so, a postelection Rutgers-Eagleton Poll found that among registered voters who said they failed to get to the polls, only 20 percent mentioned that the storm played a role.

Governor Chris Christie

Sandy's influence on the 2012 presidential election pales in comparison with how it changed the course of New Jersey politics. From inauguration to September 2012, Christie's favorability numbers hovered around the 50 percent mark in every Rutgers-Eagleton Poll, and his job grade stayed within the realm of a B or C. Yet rampant speculation over a 2012 presidential bid for Christie, followed by rumors that he might be chosen as Romney's running mate, left many New Jerseyans yearning for new state leadership. In the final Rutgers-Eagleton Poll before Sandy, 44 percent thought he deserved a second term while 47 percent said it was time for someone new.[29]

The hurricane quickly changed all of that, as shown in figure 4.3. Christie's rhetoric and actions immediately before, during, and after the hurricane, as well as his now iconic embrace with President Obama, propelled him to the top as a political force and role model of bipartisan cooperation. Christie and his famous fleece jacket were the pride of the state. In the first Rutgers-Eagleton Poll after the storm, 67 percent of New Jersey voters had a favorable impression of the governor (a jump of 19 points), and 59 percent

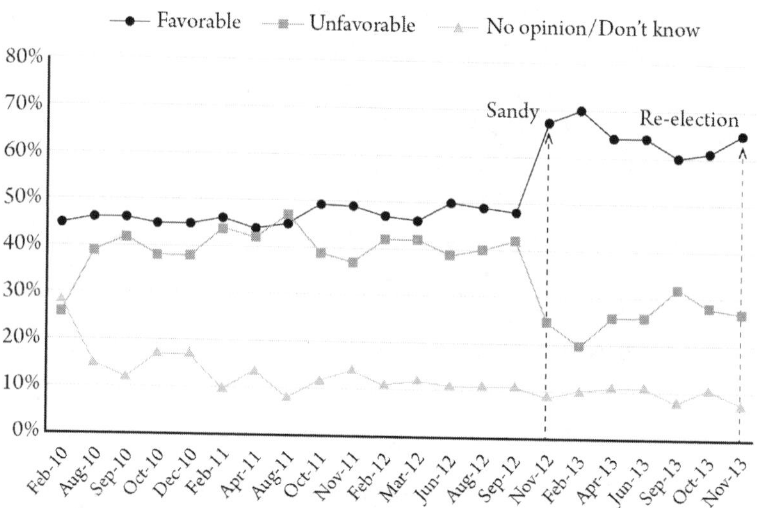

FIGURE 4.3. Question asked of New Jersey registered voters, unless otherwise specified. "First, I'd like to ask about some people and groups. Please tell me if your general impression of each one is favorable or unfavorable, or if you do not have an opinion: Governor Chris Christie." (Source: Rutgers-Eagleton Poll, Eagleton Center for Public Interest Polling, February 2010–November 2013.)

supported his reelection. Democrats, once his most bitter opponents, joined Republicans and independents in praising the governor's efforts.

Underlying the dramatic climb in his favorability ratings was the 61 percent of residents who said Christie's handling of Sandy made them support him more than they had before the storm hit. As figure 4.4 makes clear, Christie's Sandy-specific approval rating remained at around 80 percent or higher for the entire year leading up to his reelection, despite much lower approval in issue areas about which New Jerseyans said they cared the most, like taxes, jobs, and the economy. For these issues, his ratings were barely positive or even negative in the case of his handling of taxes. Nevertheless, few politicians in the country came close to his overall approval ratings. It is quite clear that without the Sandy-induced rally effect, Christie would have continued to be a deeply polarizing figure.

The shoot-from-the-hip and straight-talking qualities that Christie was criticized for before Sandy in poll responses and in media reports became

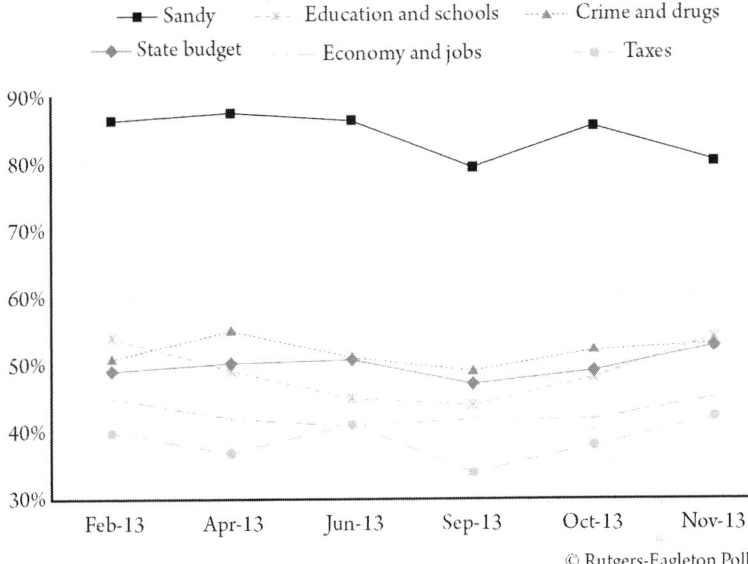

FIGURE 4.4. Question asked of New Jersey registered voters. "Now I am going to list some specific areas where I would like you to tell me if you approve or disapprove of the job Chris Christie is doing [. . .]." (Source: Rutgers-Eagleton Poll, Eagleton Center for Public Interest Polling, February 2013–November 2013.)

his biggest assets afterward, as shown in figure 4.5. Previous portrayals of Christie as rude morphed into a depiction of no-nonsense governing. Negative traits were much less likely to be applied to Christie post-Sandy; trait descriptions like "bully," "stubborn," and "arrogant" all declined. More than half of voters said Christie made them "proud," a large gain from before the storm.

Christie was now a beacon of bipartisanship, a new role for the governor that would eventually guarantee his reelection victory. Eight in ten New Jerseyans believed Christie and Obama showed "needed cooperation and bipartisanship"; three-quarters still felt this way more than half a year later.

With the gubernatorial election a year away, the only Democrat to challenge Christie was State Senator Barbara Buono, who officially announced her candidacy a little more than a month after Sandy hit.[30] Buono made little ground over the course of the campaign with her party base, party leadership, or with New Jersey as a whole. Christie maintained a double-digit lead

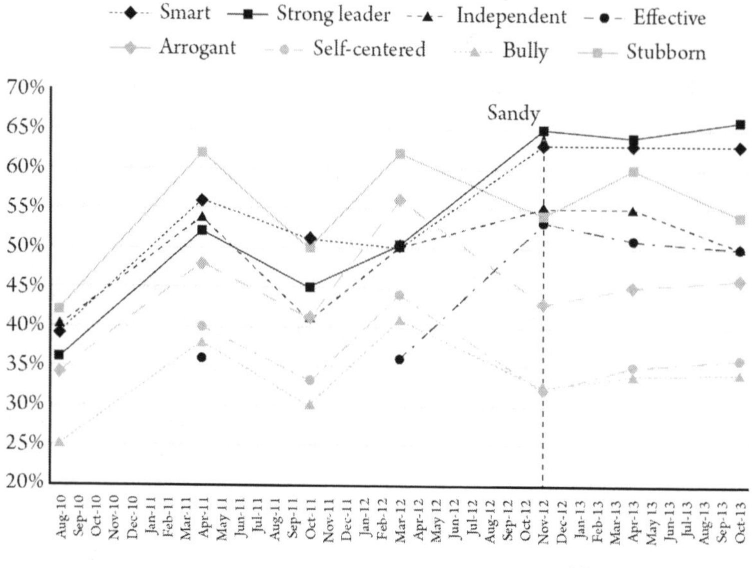

FIGURE 4.5. Question asked of New Jersey registered voters. "I am going to read a list of words that might describe Governor Christie. For each word, tell me if it describes him very well, somewhat well, or not at all." (Source: Rutgers-Eagleton Poll, Eagleton Center for Public Interest Polling, August 2010–October 2013.)

over Buono throughout the entire election season, winning by 22 points on Election Day, 60 percent to 38 percent.[31] Sandy proved to be the perfect political storm for Christie.

For a long time it seemed as if Christie's rally effect would never end and would push him inexorably toward the White House. But gravity is strong, and like all rally effects, Christie's Sandy-induced stardom did eventually disappear. His second term opened with media reports about a contrived September 2013 traffic jam at the George Washington Bridge, one of the busiest bridges in the world, orchestrated by top staff members and appointees in his administration to apparently harm a local politician, with (as yet unproven at this writing) allegations that Christie himself was involved.[32] As the scandal unfolded into legislative and U.S. Attorney investigations, the public turned on Christie, or at least Democrats did. Immediately after a January 9, 2014, press conference where Christie denied personal involvement but threw members of his staff "under the bridge," the

Rutgers-Eagleton Poll recorded a precipitous drop from 65 percent favorable to 46 percent favorable, led by a 26-point decline among Democrats. Voters were slightly more positive about Christie's performance as governor, with half approving how he handled the job, but this, too, showed a double-digit drop from November.[33] His ratings continued to plummet: by August 2015 Christie's favorability and job approval were both net negative, at 30 percent favorable to 59 percent unfavorable and 37 percent approve to 59 percent disapprove.

As additional allegations came out, this time surrounding Christie's administration withholding federal funding aid for Sandy recovery from mayors who would not endorse his reelection, Christie's Sandy-specific rating dropped dramatically as well, down to 54 points by late February 2014, a decline of 26 points from his final pre-reelection rating; this number dropped even further to 46 percent in August 2015.[34]

Christie's banner year thanks to Sandy is a fascinating study and an unprecedented instance of how a natural disaster has produced a prolonged rally effect for a politician. While natural disasters and their aftermath have often led to blame and electoral repercussions for officeholders, Hurricane Sandy did just the opposite for Christie.[35] And although an unpopular governor in New Jersey, Christie continues to draw on Hurricane Sandy to feed national political ambitions, announcing his bid for the 2016 GOP presidential nomination on June 30, 2015, and placing at the head of fundraising for his campaign the former executive director of the Hurricane Sandy Relief Fund. But what the storm has given to Christie, it has also taken away: controversies abounded regarding the administration's mishandling of the "Stronger than the Storm" tourism campaign during summer 2013, improper allocation and withholding of Sandy funds, and lack of visible progress with Sandy recovery for many homeowners who still remain homeless from the storm.[36]

LESSONS

Intensive polling after Sandy reinforces lessons about the usefulness of surveying the general public to assess diffuse experiences of a major event. Poll findings can be an important source of information for officials and other leaders in planning for and responding to disasters. Survey data can also help us to understand the attitudes and behaviors of individuals and the world around us, and can be particularly powerful during unprecedented times.

- Polling of personal reactions to posthazard events can provide a record of public intent and expectations that can be used to inform official action.

In Sandy's aftermath, surveys helped to quantify and permanently record the devastation, emotions, thoughts, hopes, future plans, and requests of citizens both affected and unaffected by the storm. Survey data have given voice to individuals' views on the hurricane itself, how it has changed their lives, and how they would like to see recovery and rebuilding proceed. These views can be used to inform official action; however, as we see here, they often did not accord with the actions that officials were taking. For example, three months after Sandy, almost two-thirds of state residents believed that rebuilding at the shore should proceed only after analysis of potential future impacts related to climate change.

- Polling after natural hazard events yields insights into the intersection of emotional effects and politics.

Two staples of polling, questions about personal experiences and political preferences, intersected in important ways after this disaster. Public opinion data showed both the personal and political sides of Hurricane Sandy and how the two overlapped, particularly in New Jersey, where Sandy completely altered perceptions of Governor Christie and assured his winning the 2013 gubernatorial election.

- Ongoing polling can help to identify whether Sandy was transformational in the long term, changing not just behavioral intentions, but actual behavior.

Survey data will also help us to track, study, and analyze longer-term trends, especially in the aftermath of events like Sandy. While we may not need a poll to tell us that Sandy was a transformative event, survey data help to tell us why this is so and in what ways. Ongoing polling by Eagleton and other pollsters makes it clear that recovery from a storm takes years for many people, and that beliefs and experiences of an extreme event, and of government responses, have lasting political effects.

NOTES

Special thanks to members of the Rutgers-Eagleton Poll Staff for their assistance with research and analysis: Caitlin Sullivan, Elizabeth Kantor, John Masusock, Ian McGeown, Gabriela Perez, and Jingying Zeng.

1. CNN, "Hurricane Sandy Fast Facts," July 13, 2013, http://www.cnn.com/2013/07/13/world/americas/hurricane-sandy-fast-facts; State of New Jersey, Governor Chris Christie, "Christie Administration Releases Total Hurricane Sandy Damage Assessment of $36.9 Billion," accessed November 28, 2012, http://www.state.nj.us/governor/news/news/552012/approved/20121128e.html; Centers for Disease Control and Prevention, "Deaths Associated with Hurricane Sandy—October–November 2012," accessed May 24, 2013, http://www.cdc.gov/mmwr/preview/mmwrhtml/mm6220a1.htm.
2. Details of all of surveys, as well as additional reports and data, are available at http://eagletonpoll.rutgers.edu.
3. Results are from a Rutgers-Eagleton Poll of 1,228 New Jersey adults contacted by live callers conducted statewide among both landline and cell phone households from November 14 to 17, 2012. The margin of error is +/- 2.8 percentage points. Within this sample is a subsample of 1,108 registered voters; this subsample has a margin of error of +/- 2.9 percentage points.
4. New Jersey is generally divided into five regions for polling: urban areas, which include Hudson and Essex counties, immediately adjacent to New York City; suburban counties, the first ring of counties outside the northeastern urban center; exurban counties in northwest New Jersey, which are somewhat more rural and in many places quite wooded; shore counties, which are immediately next to the Atlantic Ocean; and the southern part of the state, which includes the New Jersey suburbs of Philadelphia and further south.
5. Results are from a Rutgers-Eagleton Poll of 918 New Jersey adults contacted by live callers conducted statewide among both landline and cell phone households from March 8 to 17, 2013. The margin of error is +/- 3.2 percentage points.
6. Raymond Hernandez, "Congress Approves $51 Billion in Aid for Hurricane Victims," *New York Times,* January 28, 2013, http://www.nytimes.com/2013/01/29/nyregion/congress-gives-final-approval-to-hurricane-sandy-aid.html?_r=0.
7. Results are from a Rutgers-Eagleton Poll of 796 New Jersey adults contacted by live callers conducted statewide among landline and cell phone households from January 30 to February 3, 2013. The margin of error is +/- 3.5 percentage points.
8. These data are part of Monmouth University Polling Institute's Sandy recovery tracking survey. Pollsters interviewed 854 New Jersey residents who "suffered significant damage to their primary home and applied for state assistance through the 'reNew Jersey Stronger Program,'" either online or by telephone between September 18, 2013,

and January 8, 2014, as part of a larger panel study "designed to track the experiences of New Jersey residents who continue to be impacted by the storm." The study used nonprobability methods to obtain respondents and therefore cannot be generalized to "the larger population of Sandy victims in the state." This specific data can be found in the institute's February 17, 2014, report. This report and others related to the panel study can be found at http://www.monmouth.edu/university/monmouth-university-poll-reports.aspx.

9. Kelly Roncace, "'Stronger Than the Storm' Aims to Empower New Jersey, Help Those Affected by Sandy," *South Jersey Times* on NJ.com, July 8, 2013, http://www.nj.com/indulge/index.ssf/2013/07/stronger_than_the_storm_aims_to_empower_new_jersey_help_those_affected.html.

10. Jenna Portnoy and MaryAnn Spoto, "Christie, Celebs Mark Summer's Start with Sandy Recovery Themed 'Today' Show in Seaside Heights," *Star-Ledger* on NJ.com, May 24, 2013, http://www.nj.com/politics/index.ssf/2013/05/christie_today_show_seaside_he.html.

11. Results are from a Rutgers-Eagleton Poll of 888 New Jersey adults contacted by live callers conducted statewide among both landline and cell phone households from June 3 to 9, 2013. The margin of error is +/- 3.3 percentage points.

12. Results are from a Rutgers-Eagleton Poll of 804 New Jersey registered voters contacted by live callers conducted statewide among both landlines and cell phones from October 28 to November 2, 2013. The margin of error is +/- 3.4 percentage points.

13. Results are from a Rutgers-Eagleton Poll of 923 New Jersey adults contacted by live callers conducted statewide among both landline and cell phone households from April 3 to 7, 2013. The margin of error is +/-3.2 percentage points.

14. Anthony Leiserowitz, Edward Maibach, Connie Roser-Renouf, Geoff Feinberg, Seth Rosenthal, Jennifer Marlon, and Peter Howe, *Extreme Weather and Climate Change in the American Mind*, Yale University and George Mason University (New Haven, CT: Yale Project on Climate Change Communication), November 2013, http://environment.yale.edu/climate-communication/files/Extreme-Weather-Public-Opinion-November-2013.pdf.

15. Results are from a Rutgers-Eagleton Poll of 923 New Jersey adults contacted by live callers conducted statewide among both landline and cell phone households from April 3 to 7, 2013. The margin of error is +/-3.2 percentage points.

16. Results are from a Rutgers-Eagleton Poll of 842 New Jersey adults contacted by live callers conducted statewide among both landline and cell phone households from February 22 to 28, 2014. The margin of error is +/- 3.7 percentage points. Within this sample are 729 registered voters, with a margin of error of +/- 3.8 percentage points.

17. Brian Thompson, "New Jersey Town Builds Dikes to Sandy-Proof Its Neighborhoods," NBC New York Local, March 31, 2014, http://www.nbcnewyork.com/news/local/Secaucus-Builds-Dikes-Sandy-Proof-Low-Lying-Neighborhoods-253271771.html; Wayne Parry, "After Sandy, Feds Consider Building Fake Islands off the Coast," *Huffington Post*, March 29, 2014, http://www.huffingtonpost.com/2014/03/29/sandy-artificial-islands_n_5054956.html?utm_hp_ref=climate-change; Sarah Watson, "Experts: Rebuild from Sandy with Climate Change in Mind," *Press of Atlantic City*,

October 15, 2013, http://www.pressofatlanticcity.com/news/press/atlantic/experts-rebuild-from-sandy-with-climate-change-in-mind/article_a414ab5c-812c-5685-9b80-4a5188112994.html.

18. These data are part of Monmouth University Polling Institute's Sandy recovery tracking survey. Pollsters interviewed 1,239 New Jersey residents who "suffered significant damage to their primary home (defined as having more than one foot of water in the first floor or at least $8,000 in property damage)," either online or by telephone between September 18, 2013, and January 8, 2014, as part of a larger panel study "designed to track the experiences of New Jersey residents who continue to be impacted by the storm." The study used nonprobability methods to obtain respondents and therefore cannot be generalized to "the larger population of Sandy victims in the state." The specific data can be found in the institute's March 20, 2014, report. This report and others related to the panel study can be found at http://www.monmouth.edu/university/monmouth-university-poll-reports.aspx.

19. Gregory Holyk, Seth Brohinsky, Dean Williams, Damla Ergun, Gary Langer, and Julie Phelan, "Polling in the midst of a Natural Disaster: The ABC News/Washington Post 2012 Election Tracking Poll and Hurricane Sandy," paper presented at the 68th annual conference of the American Association for Public Opinion Research, Boston, May 16–19, 2013.

20. Ibid.; Josh Jordan, "Polling Wrap-Up and Hurricane Sandy's Impact," *National Review*, October 29, 2012, http://www.nationalreview.com/corner/331987/polling-wrap-and-hurricane-sandys-impact-josh-jordan; Jonathan Easley, "Sandy to Suspend Gallup and Make Polls Even More Confusing," *Hill*, October 30, 2012, http://thehill.com/blogs/blog-briefing-room/news/264785-sandys-havoc-to-make-polls-even-more-confusing; Mark Blumenthal, "Presidential Polls Remain Steady, but Will Hurricane Sandy Disrupt Tracking?," *Huffington Post*, October 29, 2012, http://www.huffingtonpost.com/2012/10/29/presidential-polls-hurricane-sandy_n_2038251.html; Brett Logiurato, "Gallup Suspends Polling Due to Hurricane Sandy," *Business Insider*, October 29, 2012, http://www.businessinsider.com/gallup-suspends-polling-hurricane-sandy-daily-tracking-obama-romney-2012-10.

21. CNN, "Obama, Romney Cancel Campaign Events ahead of Hurricane Sandy," PoliticalTicker, October 27, 2012, http://politicalticker.blogs.cnn.com/2012/10/27/obama-cancels-some-campaign-events-ahead-of-hurricane-sandy.

22. Mary Bruce, "Obama Visits FEMA HQ before Touring Storm Damage," ABC News, October 31, 2012, http://abcnews.go.com/blogs/politics/2012/10/obama-visits-fema-hq-before-touring-storm-damage; Ewen MacAskill, "Obama Tours Shattered New Jersey and Promises Help for Sandy Victims," *Guardian*, October 30, 2012, http://www.theguardian.com/world/2012/oct/31/obama-new-jersey-sandy-victims.

23. Stephanie Condon, "How Will Sandy Impact Chris Christie's Political Future?" CBS News, October 31, 2012, http://www.cbsnews.com/news/how-will-sandy-impact-chris-christies-political-future; Maureen Dowd, "The 'I' of the Storm," *New York Times*, October 30, 2012, http://www.nytimes.com/2012/10/31/opinion/dowd-the-i-of-the-storm.html; Jon Ward, "Chris Christie Denied Mitt Romney Request to Appear at Campaign Event Days ahead of 2012 Election," *Huffington Post*, November 5, 2012, http://www.huffingtonpost.com/2012/11/05/chris-christie-mitt-romney_n_2079371.html.

24. Nate Silver, "Impact of Hurricane Sandy on Election Is Uncertain," FiveThirtyEight, *New York Times*, October 29, 2012, http://fivethirtyeight.blogs.nytimes.com/2012/10/29/impact-of-hurricane-sandy-on-election-is-uncertain; Harry J. Enten, "Was It Hurricane Sandy That Won It for President Obama?," *Guardian*, December 4, 2012, http://www.theguardian.com/commentisfree/2012/dec/04/hurricane-sandy-won-president-obama; CBS, "What Impact Did Superstorm Sandy Have on the Election?," November 7, 2012, http://newyork.cbslocal.com/2012/11/07/what-impact-did-superstorm-sandy-have-on-the-election/; Real Clear Politics' 2012 general election average that aggregated all major national popular vote polls throughout the campaign can be found at http://www.realclearpolitics.com/epolls/2012/president/us/general_election_romney_vs_obama-1171.html.

25. Paige Lavender, "Mitt Romney: Hurricane Sandy 'Didn't Come at the Right Time,'" *Huffington Post*, June 6, 2013, http://www.huffingtonpost.com/2013/06/07/mitt-romney-hurricane-sandy_n_3404889.html.

26. Yamil Velez and David Martin, "Sandy the Rainmaker: The Electoral Impact of a Super Storm," *PS: Political Science and Politics* 46, no. 2 (2013): 313–323.

27. Christopher Baxter, "N.J. Sees Record-Low Voter Turnout in Wake of Hurricane Sandy," *Star-Ledger* on NJ.com, November 7, 2012, http://www.nj.com/politics/index.ssf/2012/11/nj_sees_record-low_voter_turno.html.

28. Ibid.

29. Results are from a Rutgers-Eagleton Poll of 790 New Jersey registered voters contacted by live callers conducted statewide among both landline and cell phone households from September 27 to 30, 2012. The margin of error is +/-3.5 percentage points.

30. John Celock, "Barbara Buono Set to Announce Chris Christie Challenge in 2013 New Jersey Governor Race," *Huffington Post*, December 11, 2012, http://www.huffingtonpost.com/2012/12/11/barbara-buono-chris-christie-governor_n_2276289.html.

31. Real Clear Politics' 2013 New Jersey gubernatorial election average and the election's actual outcome can be found at http://www.realclearpolitics.com/epolls/2013/governor/nj/new_jersey_governor_christie_vs_buono-3411.html.

32. Darryl Isherwood, "10 Things You Need to Know about the George Washington Bridge Scandal," NJ.com, January 9, 2014, http://www.nj.com/politics/index.ssf/2014/01/10_things_you_absolutely_need_to_know_about_the_george_washington_bridge_scandal.html.

33. Results are from a Rutgers-Eagleton Poll of 826 New Jersey adults contacted by live callers conducted statewide among both landline and cell phone households from January 14 to 19, 2014. The margin of error is +/-3.4 percentage points. Within this sample are 757 registered voters, with a margin of error of +/- 3.6 percentage points.

34. Leigh Ann Caldwell, Chris Frates, Steve Kastenbaum, and Cassie Spodak, "Hoboken Mayor: 'It's True' Christie Administration Withheld Sandy Funds," CNN, January 20, 2014, http://www.cnn.com/2014/01/18/politics/hoboken-mayor-christie-sandy-funds. Results are from a Rutgers-Eagleton Poll of 757 New Jersey registered

voters contacted by live callers conducted statewide among both landline and cell phone households from July 25 to August 1, 2015. The margin of error is +/-4.0 percentage points.

35. Christopher H. Achen and Larry M. Bartels, "Blind Retrospection: Electoral Responses to Drought, Flu, and Shark Attacks" (paper presented at the annual meeting of the American Political Science Association, Boston, 2002); Achen and Bartels, "Blind Retrospection: Why Shark Attacks Are Bad for Democracy," Center for the Study of Democratic Institutions (working paper, Vanderbilt University, 2013); F. Glenn Abney and Larry B. Hill, "Natural Disasters as a Political Variable: The Effect of a Hurricane on an Urban Election," *American Political Science Review* 60 (1966): 974–981; Kevin Arceneaux and Robert Stein, "Who Is Held Responsible When Disaster Strikes? The Attribution of Responsibility for a Natural Disaster in an Urban Election," *Journal of Urban Affairs* 28 (2006): 43–53; John T. Gasper and Andrew Reeves, "Make It Rain? Retrospection and the Attentive Electorate in the Context of Natural Disasters," *American Journal of Political Science* 55 (2011): 340–355; Andrew Healy and Neil Malhotra, "Random Events, Economic Losses, and Retrospective Voting: Implications for Democratic Competence," *Quarterly Journal of Political Science* 5, no. 2 (2010): 193–208; Ashley Koning and David P. Redlawsk, "Rally 'round the Governor: The Response of Voters to Gubernatorial Leadership in Times of Crisis," in *The American Governor: Power, Constraint, and Leadership in the States*, ed. David P. Redlawsk (New York: Palgrave Macmillan, 2015), 177–198.

36. Shushannah Walshe, "Gov. Chris Christie Faces Yet Another Controversy over Sandy Ads," ABC News, January 13, 2014, http://abcnews.go.com/blogs/politics/2014/01/gov-chris-christie-faces-yet-another-controversy-over-sandy-ads; Leslie Larson and Dan Friedman, "Stronger Than the Storm? Gov. Chris Christie Faces Federal Investigation over Sandy Recovery Ads Featuring His Family during Re-election Campaign," *New York Daily News*, January 13, 2014, http://www.nydailynews.com/news/politics/christie-faces-federal-investigation-sandy-tv-ads-article-1.1577778; ABC Local News, "Sandy Aid Distribution Mishandled, Critics Say," February 11, 2014, http://abclocal.go.com/wpvi/story?id=9427406; Matt Friedman, "Christie Used Sandy Funds for Senior Complex in Town Where Mayor Endorsed Him," *Star-Ledger* on NJ.com, January 28, 2014, http://www.nj.com/politics/index.ssf/2014/01/questions_raised_about_christies_use_of_sandy_funds_to_build_complex_in_town_where_mayor_endorsed_hi.html; Tracey Samuelson, "17 Months Later, Some Sandy Victims Still Waiting to Go Home," WNYC News, March 28, 2014, http://www.wnyc.org/story/eighteen-months-later-some-sandy-victims-still-waiting-go-home.

5 • ECOLOGICAL INJURY AND RESPONSES TO HURRICANE SANDY

Physical Damage, Avian and Food Web Responses, and Anthropogenic Attempts to Aid Ecosystem Recovery in New Jersey Estuaries

JOANNA BURGER AND LARRY NILES

Much of the attention in the aftermath of Hurricane Sandy has concentrated on the recovery, preparedness, and resiliency of coastal communities, including the importance of soft barriers or infrastructure in protecting human communities. Soft infrastructure for shorelines usually refers to sand bars, barrier islands, beaches, dunes, and salt marshes. When these features develop naturally, they become integral elements of specialized ecosystems that mediate between sea and land. The species that depend on these landscape elements also affect how sand and silt accumulate, especially vegetation species. Most members of the public are more familiar with hard barriers or infrastructure, such as sea walls, than with soft barriers. While managers have long recognized the importance of coastal zone management and soft infrastructure for human communities, Sandy dramatically brought this lesson home to people of the Northeast.[1]

The recovery and resiliency of the ecological systems underpinning our coastal communities has received considerably less attention. The resiliency of ecological systems can be examined within the context of human communities or within an ecological framework that focuses on ecosystem structure and function, with people as one forcing function, albeit a very important one because of our disruption of the coastal zone.

In this chapter we describe the ecological damages to New Jersey's major bays and estuaries, explore the storm's effects on the organisms that live in these habitats, and discuss the natural and anthropogenic responses to ecological damages from this severe storm. We were particularly interested in the interplay between physical damage, avian and food web responses, and human responses to restoration and recovery of biological resources. We focus on birds for several reasons. Their presence and occurrence is more obvious than some other species groups; the major bays and estuaries of New Jersey have abundant and diverse nesting communities of birds; many birds are high on the food web (and thus may reflect contamination); and lastly, birds are of interest to many people. We discuss the three main estuaries in New Jersey, the New York–New Jersey Harbor Estuary, Barnegat Bay, and Delaware Bay.

Coastal New Jersey and adjacent New York took the brunt of Sandy because it stalled over the region, resulting in more severe coastal damage, greater surge and tidal flooding, and more overwash of sensitive beaches with associated removal of sand. Because of the intense coastal development in the region over several decades, natural ecological structures are no longer intact as a system, and a variety of geological and ecological processes have been inhibited, thereby reducing natural protections along much of the coastline. Humans have added hard and soft infrastructure along portions of the coast, and this appears to have reduced damage in some places, but artificial dunes and other barriers were often overwhelmed. The range of damage discussed here is typical of the coastal areas that were directly hit by Sandy, although at the high end.

FRAMEWORK FOR UNDERSTANDING ESTUARINE ECOSYSTEM RESPONSES

Understanding ecosystem responses requires assessing physical aspects, species responses and interactions, anthropogenic drivers of coastal vulnerability, natural ecosystem recovery, and anthropogenic restoration. Coastal ecologies are routinely shaped by storms. Sandy did transform natural systems in some ways, although the most worrisome changes were to sites that had been previously altered by human activities in ways that already reduced their ecological functions. We suggest that Sandy also had a transformational effect on how managers, funding sources, and the public responded to ecosystem vulnerability, recovery, and restoration. Following Sandy, there was a much greater awareness of the importance of multidisciplinary, interdisciplinary, interagency, and interorganizational collaboration to respond immediately, and for longer durations, to restore and increase resiliency of coastal ecosystems, for themselves, as well as for human communities.

We make the following eight arguments:

1. The New Jersey coast has been altered and buttressed against storms for many years by the U.S. Army Corps of Engineers.
2. Physical disruptions to coastal margins can be natural or anthropogenic.
3. Physical disruptions can be measured.
4. Physical disruptions can be restored, although the efficacy varies both spatially and temporally.
5. Physical disruptions affect biota (plants and animals) within these ecosystems.
6. Recovery and resilience of biota may depend on both physical resiliency (dunes, beaches, and the soil basis of salt marshes), and on the presence and resiliency of lower trophic levels (some birds depend on marsh invertebrates, while others depend on a fish prey base).
7. Recovery and resilience of biota may be enhanced by anthropogenic actions, and the time lag to recovery may be shortened by these actions.
8. Restoration or recovery of ecosystem structures to protect human resources (e.g., sandy beaches for tourism or bulkheads for marina protection) depends partly on established agencies (e.g., the U.S. Army Corps of Engineers and the Federal Emergency Management Administration).

Restoration for enhancing the temporal and spatial recovery of ecosystem structure may require more collaboration among agencies, individuals, nongovernmental groups, and interest groups, and is therefore likely to have less access to steady funding and organizational support.

These aspects will be discussed within the context of the three estuaries, and the potential effect on birds and the food web. In each estuary, we discuss the birds that are of most conservation concern for that region and the human responses to damage from Sandy. We focus most of our attention on Delaware Bay, where Sandy resulted in the greatest change in ecological management and restoration. Severe flood surges could result in different potential effects. Using contaminants as an example of ecosystem effects: (1) storm surges could scour the bottom, taking out sediment (which could either uncover contaminants, or remove soil with contaminants); (2) storm surges could cover up contaminants with oceanic sand, reducing exposure for biota; or (3) storm surges could go up-river and carry contaminants down-river to the estuary when they flow downstream. We caution that the effects of Sandy must be considered within the context of the extensive modifications and protections of the Army Corps. While the agency's beach nourishment has helped many beach communities, the process and physical structures along the beach have prevented the natural movement of sand.

Another important component is existing degradation at the time of Sandy, which varied among the three estuaries (table 5.1). Development and degradation was extensive in the New York–New Jersey Harbor Estuary prior to Sandy, obliterating most natural shorelines. In Barnegat Bay, degradation and development of the beaches on the barrier islands was extensive, with some beach nourishment undertaken by the Army Corps of Engineers on the ocean side of the islands, but the marshes adjacent to the barrier islands and mainland were largely undisturbed, except for extensive mosquito ditching and open water marsh management in some places. Delaware Bay is the most undeveloped in terms of cities and urbanization, with low populations along the shores, but the marshes have suffered from a lack of management. Old, abandoned salt hay farms (*Spartina* sp.) have not been restored to a natural salt marsh condition. Salt marshes have been disappearing in the Northeast for some time.[2] Sandy's impacts operated on top of this development and degradation.

TABLE 5.1 Comparison of the Three Major New Jersey Estuaries with Respect to Physical Aspects, Physical Damage, Human Communities, Birds, Food Web Responses, and Anthropogenic Responses

	NY-NJ Harbor Estuary	Barnegat Bay	Delaware Bay
Physical aspects	Open bay with large rivers entering (Hudson River, Raritan River), few islands in the main bay, but many islands in the tributary rivers.	Open linear bay bordered by barrier islands and the mainland. Salt marshes fringe both east and west sides of the bay. Over 300 small salt marsh islands within the bay.	The Delaware River enters this large, open bay. There are no barrier islands or small islands. Beach habitat is intermixed with salt marshes and small rivers entering the bay.
Overall level of ecological degradation	Very high, with extensive bulkheading and development along most water and land interfaces.	High on barrier islands; intermediate to high on mainland side, with some extensive marshes remaining; mosquito ditching has resulted in some marsh degradation.	Intermediate along many beaches; higher on marshes, as abandoned salt hay farms have left the marshes lower than normal, causing erosion of marshes and beaches. High on Delaware River; failed or failing towns have left miles of shoreline with damaging riprap.
Human communities and development	Highly developed, industrial, commercial, and residential settlements.	Beach communities on the barrier islands; established communities on the mainland.	Very small and scattered communities, many primarily with summer residents.

(*continued*)

Physical damage from Sandy	Extensive damage to leading land edge and to Sandy Hook. Extensive flood surges in ecological (especially Sandy Hook) and human communities. Floodwaters remained in inland areas.	Extensive damage to beach, barrier islands, and mainland; little damage to salt marshes. Flood surge left ocean water in Barnegat Bay, especially in the north without a permanent inlet.	Complete removal of sand from many beaches, leaving mudflats and rubble. Destruction of small bay communities.
Avian and food web responses	Colonial birds nest on Sandy Hook and riverine islands. Little destruction to nesting habitat for colonial birds was apparent, although beach-nesting birds had severe habitat loss (e.g., Piping Plover, Least Tern). Wash-over of salt marshes injured insect and other invertebrate prey base for birds.	Colonial birds nest extensively on salt marsh islands, which suffered little damage. But Piping Plover and Least Tern habitat on the barrier islands beaches were damaged. Prey fish less available for birds to eat the year after Sandy. Insect prey in barrier island forests less available, affecting the presence and stop-over times of spring migrants.	Key locations for migrating shorebirds that rely on Horseshoe Crab eggs for food were destroyed or rendered less suitable. Salt marsh habitats mainly unaffected, except for wash-over effects on invertebrates and insect prey base.

(*continued*)

TABLE 5.1 Comparison of the Three Major New Jersey Estuaries with Respect to Physical Aspects, Physical Damage, Human Communities, Birds, Food Web Responses, and Anthropogenic Responses (*continued*)

	NY-NJ Harbor Estuary	Barnegat Bay	Delaware Bay
Anthropogenic responses	Main concern was for the dense and devastated human communities. Main ecological concern was for Sandy Hook and associated resources.	Mainly a human-dominated response because of devastation of tourist facilities and both primary and second-home residents. Restoration of beach habitat mainly a by-product of human community needs.	Some concern for small bay communities, but main concern was for beaches destroyed by Sandy because of looming season for spawning Horseshoe Crabs and foraging shorebirds. Protection of shore communities mainly a by-product of multiorganizational protection of ecological resources through beach replenishment.
Problems with recovery and resiliency	Human needs were viewed as the major problem. Inability to work together for ecological restoration because of bi-state and local jurisdictions (home rule).	Main concern for restoration of barrier island beaches for tourism had the by-product of restoring habitat for beach nesting birds; little restoration required for salt marsh islands.	Restoration of shorebirds/crabs seen as a major problem; home-rule problems overcome by multidisciplinary, multiagency collaboration to face a common goal that was time dependent. Additional problems resulted from lack of recovery funds (some affected bay communities did not receive Sandy funding because of the low % of damage in the contiguous area).

NEW YORK–NEW JERSEY HARBOR ESTUARY

The New York–New Jersey Harbor Estuary is large and complex, with the Hudson River, Arthur Kill, Newark Bay, and Raritan River entering the Lower New York Bay and Raritan Bay. Waters from Long Island Sound and Jamaica Bay influence the ecosystem as well. Settled areas adjacent to the New York–New Jersey Harbor include New York City and sections of New Jersey that are among the most densely populated areas in the world (figure 5.1). Intensive shipping and other harbor activities mean that the harbor is deep, busy, and polluted. With so many rivers, bays, and other water bodies, water can move quickly downstream from heavy rains or upstream from ocean flooding. In addition, sea level rise has the potential to inundate islands and coastal habitats and exacerbate storm surge impacts.[3]

The harbor is home to more than four thousand colonial waterbirds of twelve species with nesting colonies on seventeen of the harbor's nineteen undeveloped islands. The birds feed in the surrounding waters, and most species either arrive well before egg-laying, or are resident.[4]

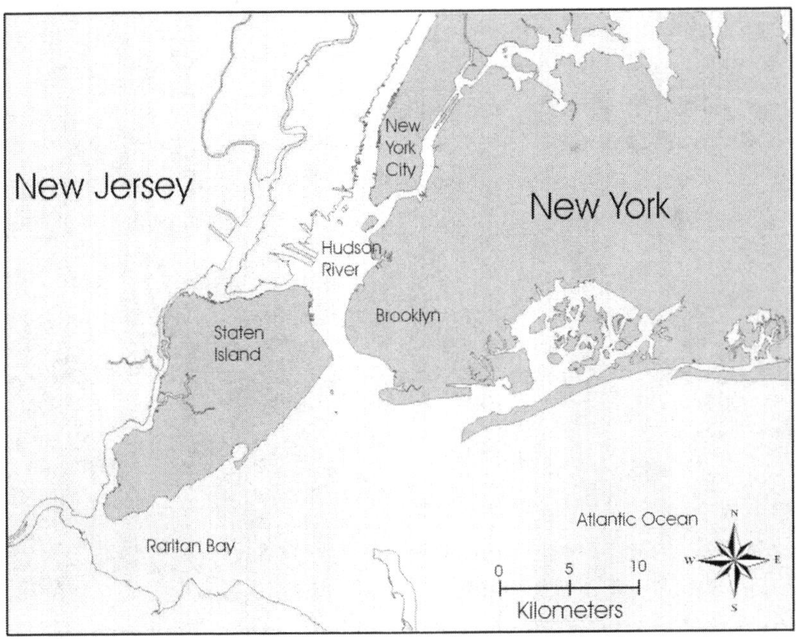

FIGURE 5.1. Map of New York–New Jersey Harbor Estuary.

Typical species feed at different levels of the food web, such as Canada Goose (*Branta canadensis*, lowest trophic level), Herring Gull (*Larus argentatus*), Great Egret (*Ardea alba*), Great Black-backed Gull (*Larus marinus*), Double-crested Cormorant (*Phalacrocorax auritus*), and Black-crowned Night-Heron (*Nycticorax nycticorax*, highest level). The colonial birds are particularly of interest because of their diversity, numbers, and appeal to the public.

There was massive flooding of ecosystems and human communities from Sandy, and elevational differences, many caused by human development, resulted in water being trapped within communities and ecosystems. Damages to leading land edges and to Sandy Hook were severe because of the strong storm surge and backwash (table 5.1). Associated damages to some birds and the food web were high, and included loss of nesting habitat for colonial birds on Sandy Hook, loss of nesting habitat for the endangered Piping Plover (*Charadrius melodius*), and loss of the food and prey base for migrant passerines at Sandy Hook and elsewhere. There were no major changes in the islands within the rivers that serve as rookeries for the nesting herons, egrets, ibises, and gulls.

The New York–New Jersey Harbor Estuary shelters large colonies of birds, including Herring Gulls (figure 5.2). Preliminary data from Herring Gull eggs indicate that levels of most heavy metals (cadmium, chromium, mercury, selenium) were lower following Sandy than in the year before. However, long-term data sets indicate that the levels following Sandy were similar to those in the 1980s and 1990s, but that levels had increased following Hurricane Irene in 2011.[5] This provides the intriguing suggestion that

FIGURE 5.2. Herring Gulls are excellent indicators of environmental health, including human disturbance, habitat loss, and contaminants.

Irene (a land-based hurricane with heavy rains that resulted in floodwaters moving down watersheds toward the New York–New Jersey Harbor Estuary) resulted in increased exposure of birds to contaminants. Hurricane Sandy may have resulted in either a covering of these contaminated soils, or a removal of them. In contrast, lead levels after Sandy were 244 ppb, as compared to pre-Sandy levels of 139 ppb. Increased lead levels may be due to receding surge floodwaters from neighborhoods with lead remaining in soil from leaded gasoline.

The prey base for migrant birds was disrupted because complete overwash of low islands removes insect prey. The islands themselves did not wash away because of the nature of the islands' construction. Most were built from dredge spoil, and many have riprap or rocky shores. In other cases, the storm surge that swept over them was deep enough so that the main force did not damage habitat (that is, it washed well above the marsh, rather than eroding the islands).

The main concern about recovery in the New York–New Jersey Harbor Estuary was for human communities because of the great destruction of homes and businesses, disruption of infrastructure, and general devastation of communities, which lasted for many months. Ecological concern focused on beach, salt marsh, and other ecological barriers that protect these human communities. Some attention to restoration of Sandy Hook and associated beaches provided habitat for nesting birds, but this response was largely motivated by protecting parkland for tourism and was driven by the U.S. Army Corps of Engineers, which does not have a specific legal requirement to address habitat enhancement. Only recently has the Corps begun to incorporate into its projects the habitat needs of birds, especially endangered species. The response with respect to ecosystem structure and function was therefore conducted in the usual manner.

BARNEGAT BAY

Barnegat Bay and Little Egg Harbor Estuary are narrow and relatively shallow in many places, with sensitive plankton communities, seagrass meadows, shellfish beds, finfish spawning habitats, nesting colonial birds, and massive numbers of wintering waterfowl.[6] The bay is bordered on the east by the two barrier islands of Island Beach and Long Beach Island, and on the west by salt marshes and the mainland (figure 5.3). There are more than 350 small islands in the bay. Many of these are overwashed by summer or winter storm tides. The washover by winter tides results in a lack of mammalian predators (any that reach the islands are washed away), making the islands ideal for nesting birds.[7]

The ocean-facing beaches on the barrier islands are home to nesting individuals of the federally endangered Piping Plover (solitary-nester), as well as state-listed endangered Least Terns (*Sterna antillarum*) and Black Skimmers (*Rynchops niger*), both of which nest in colonies. The salt marsh islands in the bay provide nesting habitat to a number of solitary nesting species, such as Willet (*Tringa semipalmata*), American Oystercatcher (*Haematopus palliatus*), Saltmarsh Sparrows (*Ammodramus* sp.), and ducks (Mallard, *Anas platyrhynchos*; Black Duck, *Anas rubripes*; and Gadwall

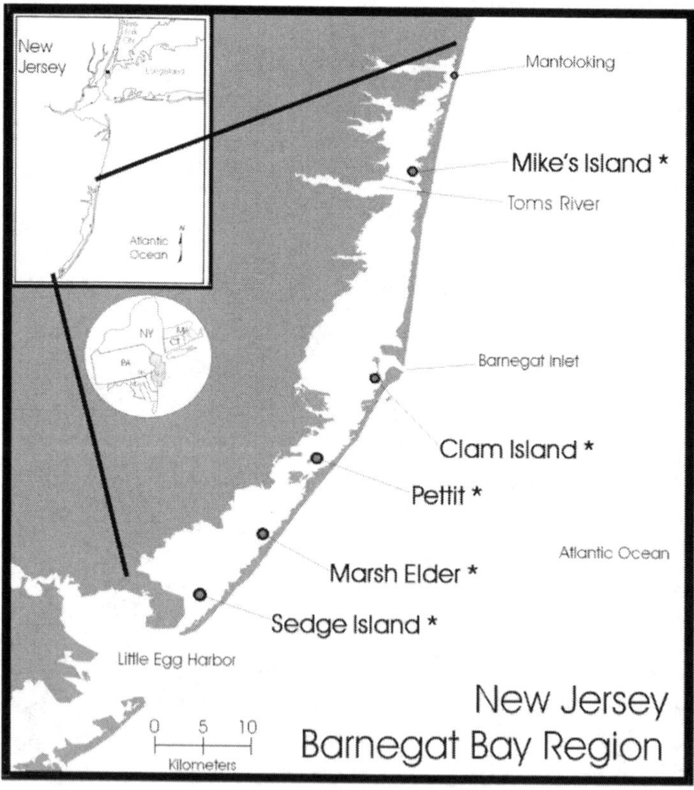

FIGURE 5.3. Map of Barnegat Bay.

Anas strepera). A large diversity of colonially nesting birds use these islands, including terns, gulls, skimmers, herons, egrets, ibises, and night herons.[8] Both the number of birds and the number of colonies has been declining over the past thirty-five years, largely owing to loss of available habitat from subsidence and sea level rise, but the changes were greater following Sandy (table 5.2).

Birds in Barnegat Bay, and elsewhere, separate by habitat type, although there is considerable overlap (figure 5.4). With increased flood surges, increased tidal range, and sea level rise, some birds will benefit, and others will not. Species that require higher nest sites, such as some gulls and some sparrows, will likely lose nesting habitat, while others that are more generalists will profit. Even so, excessively high waters will wash over salt marsh islands, washing out eggs and nests.

Sandy affected beach-nesting birds because of the washover, movement of sand, destruction of beaches and dunes, and changes in prey availability owing to the loss of some back-beach pools. Sandy did not appear to directly damage the size, shape, or elevation of the salt marsh islands where birds nested, so a similar population of birds returned to attempt nesting after Sandy. However, reproductive success was very low on the salt marsh islands. Birds attempted to nest, and even laid eggs, but not as many nests were initiated, and birds that laid eggs lost most of them to tidal flooding

TABLE 5.2 Common Terns as an Example of Changes in Numbers of Colonies and Pairs in Barnegat Bay, New Jersey

Years	Number of colonies	Number of nesting pairs
1976–1979	24	1,124
1980–1984	19	718
1985–1989	19	1,545
1990–1994	16	930
1995–1999	14	664
2000–2004	10	729
2005–2009	11	960
2010–2012	10	975
2013 (after Sandy)	6	435

SOURCES: Joanna Burger, C. David Jenkins Jr., Fred Lesser, and Michael Gochfeld, "Status and Trends of Colonially-Nesting Birds in Barnegat Bay," *Journal of Coastal Research* 32 (2001): 197–211; J. Burger, unpublished data.

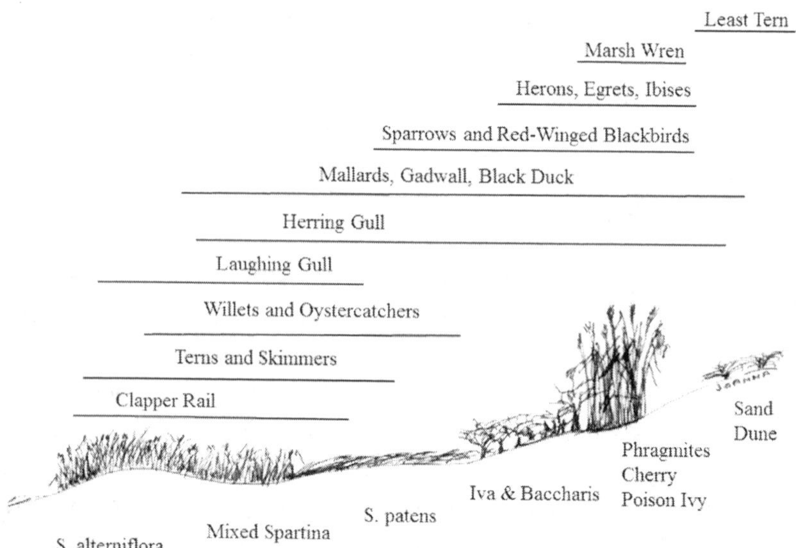

FIGURE 5.4. Schematic of nesting habitat for birds in salt marshes in Barnegat Bay. The zonation means that species of birds that are nesting lower on the marsh are more vulnerable to severe storms and sea level rise than others nesting higher or in small shrubs.

and heavy rains. Whether these losses were partly due to changes in prey availability is unclear, although species such as herons, egrets, and night-herons that partly depend on fiddler crabs may have experienced changes in prey abundance. Water had washed over these islands, removing small invertebrates.

Hurricane Sandy could have changed contaminant levels in the eggs of colonial nesting birds that feed on fish in the bays and estuaries, as well as near shore, such as Common Terns (figure 5.5). But Sandy was not followed by an increase in levels of heavy metals in eggs of Common Terns (*Sterna hirundo*) in Barnegat Bay, suggesting that the large surges of water did not shift contaminants.[9] For example, lead levels in Common Tern eggs ranged from means of 7 ppb to 25 ppb from 2008 to 2012, and was 5 ppb in 2013 (after Sandy). Over the past five years (2008–2012), mean levels of mercury ranged from 718 ppb to 1,236 ppb, and averaged 641 ppb in 2013. Similarly, levels of cadmium, the other major metal of concern in marine environments, ranged from 0.6 ppb to 13.8 ppb in the five years (2008 to 2012) before Sandy, and averaged 3.2 ppb after. The lack of a difference (pre- and

FIGURE 5.5. Common Terns that once nested on sandy barrier beaches in New Jersey have been forced to nest in salt marshes, which are vulnerable to flood tides and sea level rise.

post-Sandy) may have been due to the shallow nature of the bay, as compared with the massive surges.

Barrier-beach human communities suffered great losses, similar to those communities around the New York–New Jersey Harbor Estuary, and attention of governmental agencies mainly focused on them. However, many of the bird-nesting islands in Barnegat Bay are part of Forsythe National Wildlife Refuge, and the refuge has taken a great interest in recovery and restoration of nesting islands and in increasing their resiliency. Its efforts have been aided by the recovery funds from the U.S. Department of the Interior and by university, state agency, nongovernmental organization, and wildlife refuge collaborations to artificially increase the elevation of salt marsh islands. The aim is to decrease extreme tidal flooding and enhance habitat suitability for a suite of solitary and colonially nesting birds. Such enhancements will also increase spawning and rearing habitat for small fish, thus increasing the food base for birds. This effort is intermediate in complexity, as it involves a wide range of agencies and people from many different disciplines.

DELAWARE BAY

The Delaware River flows into the Delaware Bay, a large and complex bay that continues to widen for several miles until it reaches the Atlantic Ocean (figure 5.6). The high-volume shipping channels are maintained by dredging. The bay is bordered by a few small permanent human communities, mostly in the state of Delaware, but the majority of the bay's shoreline is

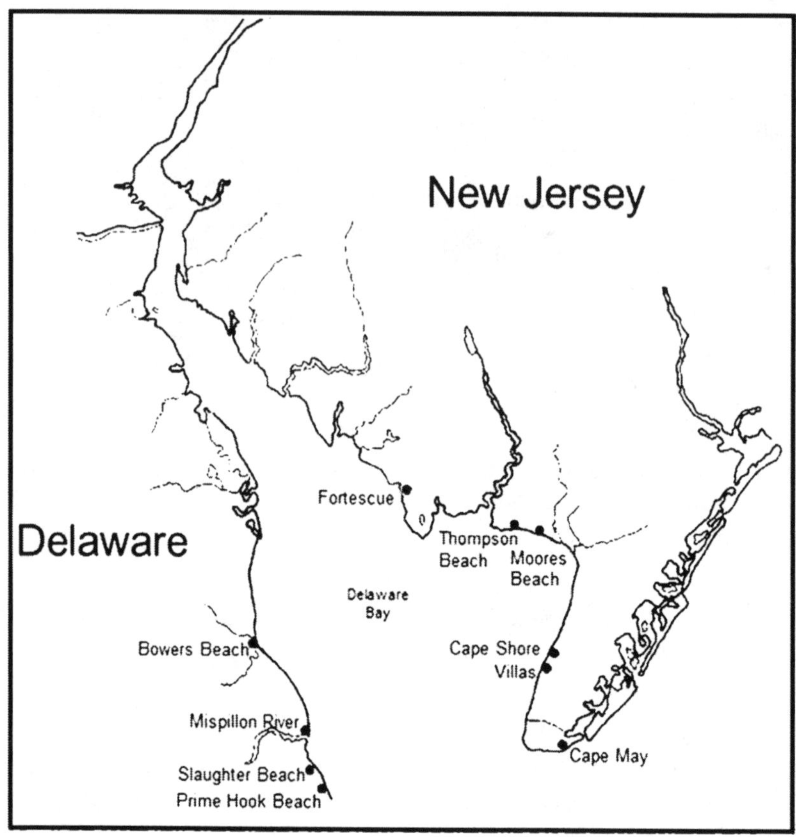

FIGURE 5.6. Map of Delaware Bay.

uninhabited in both New Jersey and Delaware. The area has a modest fishing and ecotourist economy (mostly attracting bird-watchers), but in the past, large salt hay farms dominated the coastline.

Rainfall from Sandy was extremely heavy in this area, especially along the New Jersey shore of the Delaware Bay. High winds, flooding, and flood surges resulted in damages to the human communities and in many areas completely washed most of the sand away from beaches, leaving them as mudflats or rocky beaches. In some places, beaches were littered with abandoned riprap, destroyed beach houses, furniture, and old pilings from houses long-since taken over by the bay.

Delaware Bay is bordered by sandy beaches and extensive salt marshes, with frequent streams and rivers flowing into it. Past development of marshes for salt hay farms had resulted in an overall lowering of the marshes, which may have contributed to beach erosion. The marshes provide nesting habitat for a variety of solitary-nesting birds, including Willets, Clapper Rails (*Rallus longirostris*), ducks, and sparrows. The marshes are used extensively by foraging egrets and herons that nest in nearby heronries. Since the marshes are connected to the mainland, no colonial birds nest there, because the loud, noisy colony would be obvious to predators, and would be easily accessed by mammalian predators. Laughing Gulls (*Larus atricilla*) and Herring Gulls come to forage in the spring, and they also nest in other places, such as the Stone Harbor marshes.

Delaware Bay, however, is a globally important and key migratory stopover location for shorebirds, and in the spring, more than a million shorebirds have been known to migrate through, feeding mainly on the abundance of horseshoe crab eggs (*Limulus polyphemus*) (figure 5.7).[10]

© Joanna Burger

FIGURE 5.7. Horseshoe crabs spawning on a restored beach on Delaware Bay.

While there is some controversy about the role of human overharvesting of horseshoe crabs on the availability of excess eggs for the foraging shorebirds, there is no controversy that the migrant shorebirds depend extensively on the eggs.[11] Horseshoe crabs lay their eggs below ground, where they are unavailable to the foraging birds. Eggs become available only when a superabundance of females lay on successive nights, digging up the eggs of earlier females and releasing the eggs to the surface, where the shorebirds can feed on them. The Delaware Bay supports the world's largest concentration of breeding horseshoe crabs, although their populations have declined since the 1990s.[12]

Many of the shorebird species migrate long distances to their breeding grounds, and during a few short weeks on Delaware Bay, they must nearly double their weight to make the last long flight successfully and arrive on their Arctic breeding grounds, sometimes in severe weather conditions.[13] Many of the species of shorebirds migrating through Delaware Bay have declined over the past thirty years, and they are of immense conservation concern. Red Knots (*Calidris canutus*) are particularly of concern because a significant proportion of the North American race moves through Delaware Bay, where they stop to refuel. Their numbers have declined dramatically, and they rely extensively on horseshoe crab eggs. The Red Knot has recently been placed by the U.S. Fish and Wildlife Service on the federal endangered species listing as threatened. The Knot is endangered in New Jersey and is listed as threatened in Canada.

An interdependent food web depends on the preservation of the spawning beaches. The eggs are eaten by a variety of birds, as well as by some small mammals. The small, young horseshoe crabs are eaten by a variety of fish, from killifish (*Fundulus* sp.) to weakfish (*Cynoscion regalis*), but the extent of predation is unknown. Intermediate-size crabs are eaten by larger fish as they move out of Delaware Bay onto the continental shelf.

The effects of Sandy on spawning beaches along the Delaware Bay were dramatic. Sand was removed from many beaches, exposing salt marshes to additional erosion. Habitat inventories for horseshoe crabs had indicated very little change in available habitat from 2002 to 2010. Following Sandy, though, there was a 70 percent decline in optimal habitat for spawning crabs and a 20 percent decline in suitable and less suitable habitat.[14] Optimal spawning habitat is quite nuanced, as habitat that appears suitable (for example, sandy beaches) may not have deep enough sand for the eggs to have high hatching rates. Optimal spawning beaches were the places where shorebirds also fed in abundance.

Restoring beaches is a large, time-consuming effort and requires immense quantities of sand, which is not readily available, so it called for a new level of cooperation and collaboration among management agencies (state, federal), regulators (state, federal), funding agencies (governmental, nongovernmental organizations, private), management personnel, and local landowners (many of whom had lost their beach houses). Sandy happened at the end of October, and restoration efforts had to be well organized to be done in time. Horseshoe crabs can come in to spawn in April and May; the shorebirds can arrive in early May, and they usually depart for their breeding grounds by the end of May or early June. The restoration effort therefore required putting together several commitments in a short period of time to allow the beaches to solidify enough to be appealing to the spawning crabs. Further, funders, managers and biologists had to act as one, without focusing on their own special interests.

An imposing list of organizations (table 5.3) acted to restore sand to five key beaches, restoring a total of more than a one-mile linear area. About three thousand tons of asphalt and crushed concrete were used to repair three access roads; more than eight hundred tons of rubble and other debris that could potentially trap horseshoe crabs were removed; and forty thousand tons of sand were imported.[15] While it was not possible to restore sand to all spawning beaches, or to all sections of these beaches, the effort was impressive.

The restoration efforts resulted in recreating horseshoe crab spawning areas and foraging areas for shorebirds. The numbers of horseshoe crabs and shorebirds using the restored beaches were comparable to numbers using beaches that were not damaged by Sandy and were significantly higher than at damaged beaches that were not restored (for example, control beaches; figures 5.8 and 5.9).[16] Further, the abundance of horseshoe crab eggs found both in shallow and deep locations on the restored beaches was similar to the abundance found at beaches that were left relatively undamaged by Sandy.

These data indicate that the restoration was successful, that it occurred in the necessary time frame to allow the crabs to spawn and shorebirds to feed, and that spawning rates were similar on the restored and undamaged beaches. The more remarkable aspect, however, was the collaboration that was required among so many different agencies and people, with respect to

TABLE 5.3 Resource Agencies and Funding Organizations Restoring Key Delaware Bay Horseshoe Crab Spawning Beaches

Resource/regulatory agencies	Funding organizations
U.S. Army Corps of Engineers	Eastern Partnership Office of the National Fish and Wildlife Foundation
N.J. Department of Environmental Protection, Bureau of Coastal Engineering	Geraldine R. Dodge Foundation, N.J. Recovery Fund
N.J. Department of Environmental Protection, Division of Land Use Regulation	William Penn Foundation
N.J. Department of Environmental Protection, Office of Dredging and Sediment Technology	Doris Duke Charitable Foundation
N.J. Department of Environmental Protection, Division of Fish and Wildlife	New Jersey Natural Lands Trust
N.J. Department of Environmental Protection, Commissioner's Office	Community Foundation of New Jersey
U.S. Fish and Wildlife Service, Cape May National Wildlife Refuge	Corporate Wetlands Restoration Partnership
U.S. Fish and Wildlife Service, South Jersey Field Office	American Bird Conservancy
NOAA Fisheries Service, Sustainable Fisheries Division	Cape May Mosquito Control
	The Wetlands Institute

NOTE: This effort was undertaken as a measure to improve spawning beaches and thus increase egg production so that excess eggs would be available for foraging shorebirds.

funding, design, and execution of the restoration. This was a transformational collaboration that will likely alter management actions in the future at Delaware Bay, and it provides an example for other estuaries. Success, however, will depend on continued collaboration and a regulatory framework that encourages restoration. Having established the relationships, the agencies are continuing to work together to write requests and grants to restore other beaches and to work toward enhancing trophic-level relationships within the bay. The actions required were greater than any one agency could manage (in terms of time, personnel, or funds). The estimated total cost for sand was \$3 million. The collaboration shows that it can be done and that cooperation leads to positive results for the crabs, shorebirds, other species that depend

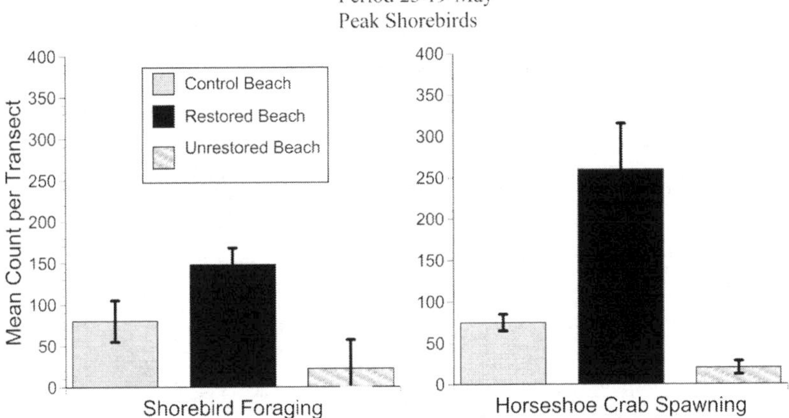

FIGURE 5.8. Comparison of shorebirds using restored beaches in 2013 compared with unrestored beaches and beaches largely unaffected by Sandy. (Source: See note 12. Data are from May 19–25, the peak stopover time for shorebirds on Delaware Bay.)

FIGURE 5.9. Shorebirds forage in dense flocks and come to forage on restored beaches because of the abundance of spawning horseshoe crabs.

on them, and for the overall food chain. And importantly, it also provided homeowners with a sandy beach under and around their houses, giving them aesthetic advantages. Moreover, it resulted in transforming animosity between different factions (bird-watchers and conservationists, crab fishermen, and homeowners) and agencies (the Army Corps of Engineers, New Jersey Department of Environmental Protection, U.S. Fish and Wildlife Service) to positive feeling about cooperating to accomplish shared goals. While animosity could have increased after a disaster such as this, after Sandy, all the stakeholders mentioned recognized that a joint effort was essential to meet a goal they all shared, ensuring the integrity of the sandy beach.

COMPARISONS OF PHYSICAL DIFFERENCES, BIOTIC RESPONSES, AND ANTHROPOGENIC COLLABORATIONS

The three major bays/estuaries are very different in terms of size and shape, physical characteristics, riverine inputs, and adjacent human communities. Their experience of the storm, and the efforts to restore ecological systems in these bays, differed as well. The New York–New Jersey Harbor Estuary and Barnegat Bay experienced relatively low rainfall but higher winds and storm surge, while the Delaware Bay experienced very high rainfall with significant winds. As described above, these differences resulted in different effects of Sandy on ecosystem structure and function, and on different anthropogenic responses to restoration of habitats, especially for birds (table 5.1).

In the New York–New Jersey Harbor Estuary, a number of different organizations (for example, New York Audubon, New Jersey Audubon, and the New York City Department of Environmental Protection) are engaged in assessment and restoration of birds and their habitats. In Barnegat Bay, the regional office of U.S. Fish and Wildlife Service, in conjunction with Forsythe National Wildlife Refuge, the National Fish and Wildlife Foundation, and university scientists, are working together to assess effects of Sandy on salt marsh habitat and birds. While these two efforts are collaborative, they do not involve as many agencies and organizations as was required to address the loss of sand on the horseshoe crab spawning beaches in Delaware Bay. The massive ecological challenge in the Delaware Bay resulted in a collaboration among agencies and organizations to restore the beaches in only five months. It transformed animosity into cooperation and respect because it was suddenly clear, and urgent, that the integrity of the beach had to be restored, and that no one group had the money, time, or expertise to do it. This joint effort restored habitat, which not only increased resiliency for beaches and associated food webs, but increased resiliency for people living along the beach, increasing their respect for this biological work.

LESSONS

There are three types of lessons learned from our comparison of ecological injury and responses to Sandy in New Jersey's major bays and estuaries and four attendant sets of recommendations. These relate to the ecological functioning of coastal systems under extreme storm surge and flooding; the interdependency of ecological and human communities; and the need for collaborations among a wide range of people and organizations.

- A storm can have widely differing effects on the ecological functions of coastal systems across a region.

Heavy surges and severe flooding had a greater effect on those areas of sandy beaches and dune habitat than on salt marsh habitat. Sand was completely removed from many beaches in Delaware Bay, and flood surges in the beach, barrier islands and mainland of the New York–New Jersey Harbor Estuary and Barnegat Bay led to extensive damage to ecological and human communities. While washover of salt marshes injured the insect and other invertebrate prey base for birds, these habitats were otherwise largely unaffected as the water moved over and above them rather than destroying them.

- Sandy revealed new lessons about the interdependency of ecological and human communities that demonstrate the value of designing restoration projects with multiple aims.

Although members of the public and local and state officials were less concerned about resiliency and recovery of ecological communities than human communities after Sandy, we see in the case of Delaware Bay that the resiliency of human communities depended partly on the resiliency of coastal ecosystems. Restoration of horseshoe crab spawning areas provided homeowners with increased recreational opportunities and aesthetic improvements. This not only suggests the importance of restoring coastal ecosystem structure and function where possible, but building from this we recommend development of targeted action plans for education and integration around resiliency and recovery.

- Transformational collaborations for recovery and resiliency can form after a disaster, even among organizations with a history of conflict. Overlaps in these organizations' values and aims and pre-storm regulatory arrangements can become the basis for cooperation. Success in restoring Delaware Bay beaches could encourage others to create advance plans.

Recovery after Sandy required collaboration among a wide range of people and organizations to accomplish immediate and timely action to restore sandy beaches. Collaboration was critical to amass the necessary funds, workforce, and equipment to act in advance of key ecological events like horseshoe crab spawning and migrant shorebird feeding. Following this, we encourage communities to develop emergency restoration plans for ecological recovery that include a wide range of nongovernmental organizations, private companies, and people in order to be able to respond immediately with money, personnel, equipment, and materials (for example, sand and gravel).

Comparing anthropogenic attempts to aid ecosystem recovery in New Jersey's three major bays and estuaries post-Sandy, we see that differences in the regulatory processes (for example, permitting) allowed for immediate recovery of key beaches and other habitats in Delaware Bay. In that ecosystem, through collaboration and streamlining, groups could address the urgency of restoration in order to meets key human and ecological needs (return to homes, tourist beaches, horseshoe crab spawning beaches). This suggests that shifts in regulatory approaches and time lines are required for timely implementation of recovery and resiliency projects.

NOTES

We want to thank the people and organizations that helped in the assessment, management, study, and restoration of the beaches along Delaware Bay. These include Joseph A. M. Smith, Dianne F. Daly, Tim Dillingham, William Shadel, Amanda D. Dey,

Mathew S. Danibel, Steven Hafner, and David Wheeler, as well as M. Gochfeld, S. Elbin, C. Jeitner, F. Lesser, T. Pittfield, and N. Tsipoura for information on the other bays. We thank the funding agencies listed in table 5.3, and Rutgers University.

1. Norbert Psuty, *Coastal Hazard Management: Lessons and Future Directions for New Jersey* (New Brunswick, NJ: Rutgers University Press, 2002).
2. New York Sea Grant, *Jamaica Bay's Disappearing Marshes: Proceedings of a Scientific Symposium and Public Forum* (New York: National Park Service and Jamaica Bay Institute, 2004).
3. New York City Panel on Climate Change, *Climate Risk Information 2013: Observations, Climate Change Projections, and Maps* (New York: New York City Mayor's Office, 2013).
4. Sharron Seitz and Stuart Miller, *The Other Islands of New York City* (Woodstock, VT: Countryman Press, 1996); Elizabeth Craig, *New York City Audubon's Harbor Heron's Project: 2013 Nesting Survey Report* (New York: New York City Audubon).
5. Joanna Burger and Susan Elbin, "Possible Hurricane Effects on Contaminant Levels in Herring Gull (*Larus argentatus*) Eggs from Colonies in the New York Harbor Complex between 2012–2013," *Exotoxicology* 249 (2015): 445–452.
6. Michael J. Kennish, "Barnegat Bay–Little Egg Harbor, New Jersey: Estuary and Watershed Assessment," *Journal of Coastal Research* 32 (2001): 3–280.
7. Joanna Burger, *A Naturalist along the Jersey Shore* (New Brunswick, NJ: Rutgers University Press, 1996); Joanna Burger, C. David Jenkins Jr., Fred Lesser, and Michael Gochfeld, "Status and Trends of Colonially-Nesting Birds in Barnegat Bay," *Journal of Coastal Research* 32 (2001): 197–211.
8. Burger et al. "Status and Trends of Colonially-Nesting Birds in Barnegat Bay."
9. Joanna Burger, unpublished data.
10. Lawrence Niles, Humphrey P. Sitters, Amanda Dey, Philip Atkinson, Alan Baker, Karen Bennett, et al., "Status of the Red Knot, *Calidris canutus rufa*, in the Western Hemisphere," *Studies in Avian Biology* 36 (2008): 1–185.
11. Lawrence Niles, Jonathan Bart, Humphry Sitters, Amanda Dey, Kathy Clark, Philip Atkinston, et al., "Effects of Horseshoe Crab Harvest in Delaware Bay on Red Knots: Are Harvest Restrictions Working?," *BioScience* 59 (2009): 153–164.
12. Lawrence Niles, Joseph Smith, Dianne Daly, Thomas Dillingham, William Shadel, Amanda Dey, Steven Hafner, and David Wheeler, "Restoration of Horseshoe Crab and Migratory Shorebird Habitat on Five Delaware Bay Beaches Damaged by Superstorm Sandy," November 22, 2013, accessed Dec. 21, 2015, http://www.smithjam.com/wp-content/uploads/2014/03/RestorationReport_112213.pdf.
13. Alan L. Baker, Patricia Gonzalez, Theunis Piersma, Lawrence Niles, Ines de Lima D. do Nascimento, Phillip Atkinson, Nigel Clark, Clive Minton, Mark Peck, and Geert Asrts, "Rapid Population Decline in Red Knots: Fitness Consequences of Refueling Rates and Late Arrival in Delaware Bay," *Proceedings of the Royal Society of London* 271 (2004): 875–882.
14. Niles et al., "Restoration of Horseshoe Crab and Migratory Shorebird Habitat."
15. Ibid.
16. Ibid.

6 • SURVIVING SANDY

Identity and Cultural Resilience in a New Jersey Fishing Community

ANGELA OBERG, JULIA A. FLAGG, PATRICIA M. CLAY, LISA L. COLBURN, AND BONNIE McCAY

Marine fishing, both commercial and recreational, contributes significantly to the economies of coastal New Jersey and New York, where Hurricane Sandy had its greatest impacts. In a letter asking Secretary of Commerce Rebecca Blank for post-storm support, Governor Chris Christie of New Jersey wrote, "In 2011, New Jersey's commercial fishing industry landed roughly 175 million pounds of seafood, generating over $1.3 billion of economic activity. Similarly, the economic impact of recreational fishing is significant, supporting, according to industry, approximately 8,500 jobs and $1.4 billion in annual sales."[1] New York had similarly important pre-Sandy levels of fishing activity. Beyond economic value, coastal communities in these states have relied on fishing for livelihoods and cultural identity going back multiple generations. Fishing contributes to the identity and character of coastal communities in ways that cannot be captured in purely monetary valuations of the industry.[2]

The question guiding this analysis is whether Hurricane Sandy was a transformational event for fishing communities in New York and New Jersey. Our research examines the cultural and economic resilience of commercial and for-hire fishermen in the communities of New York and New Jersey by using interviews to document the process of recovery in the aftermath of the storm.[3] The areas most affected by Sandy in New York and New Jersey are predominantly urban. Our research on fishing communities provides an important contribution to the growing body of post-Sandy research by recounting the experiences of people who, amid an urban landscape, are directly dependent on natural resources. Fishing communities are particularly sensitive to environmental changes, not only through the very act of fishing but also because of their proximity to the sea and dependence on viable waterfront facilities and businesses.

The impacts of Sandy provide an opportunity to observe the ways in which fishing communities recover from a shock. For the purposes of this study, we use resilience, the ability of a social system to respond to and recover from a disturbance,[4] as a lens to better understand the processes of recovery for the fishing industries in New York and New Jersey after the hurricane. The use of ethnographic interviews allowed us to assess not just resilience, but also the influence of narrative on the strength of social bonds (also called social capital); social bonds have been shown to support resilience, including in relation to disasters.[5]

DATA AND METHODS

In this study, we interviewed a total of ninety-three people in seventy-seven separate interviews in seven ports across the two states. Sixteen of these interviews were conducted in New York and the remaining sixty-one in New Jersey. Interviews in New Jersey were conducted between February 2013 and June 2013, and those in New York were conducted in September 2013. Interviewees represented several sectors within the fishing industry, including commercial fishing, party/charter recreational fishing, bait and tackle retailers, and private marinas. Loosely structured ethnographic interviews were conducted primarily in person at the ports, though nine interviews were conducted by phone.

Using an open-ended interview protocol, respondents were asked about their personal and family involvement in the fishing industry, how they prepared for the storm, what damage they saw from the storm to their home and business, how they were recovering from the damages, and what they thought the long-term impacts of the storm were. Because phone interviews tended to be shorter than in-person interviews, the interviews varied widely in duration, lasting between ten minutes and two and a half hours. Interviews were coded to identify emergent themes, and findings were analyzed across sectors and ports using Atlas.ti™, a text analysis software program.

In this chapter, we focus on the fishing cooperative (or co-op, as it is usually referred to) in Belford, Middletown Township, New Jersey, as a case study for examining potential cultural transformation. We draw on fourteen ethnographic interviews conducted at Belford, mainly in the spring of 2013, as well as observations from a preliminary visit on November 15, 2012, two weeks after the storm made landfall. Nine of the fourteen interviews were conducted with co-op members or crew from boats owned by co-op members; two were with co-op support staff; and the remaining three were with commercial fishermen who were not members of the co-op. The co-op, formed in 1953, is dominant at Belford, but some fishermen, mainly lobstermen, sell to other businesses in the port, and the co-op sometimes has nonmembers using its facilities. It is important to note there is great diversity in fishing communities and fisheries along the East Coast, including presence or absence of a cooperative. Because not all fishing communities are the same or have cooperatives, the findings from Belford may not have broad generalizability to other communities affected by Hurricane Sandy.

We highlight Belford and its cooperative for three reasons. First, Belford is the site of one of only two fishing cooperatives in New Jersey, allowing us to investigate the role of such an institution in crises. Elkeand Herrfahrdt-Pähle and Claudi Pahl-Wostl note that institutions can connect social and ecological systems and, where functional, generally serve to help maintain the status quo or at most allow incremental change.[6] However, Marie-Caroline Badjeck et al. emphasize the importance of cross-scale linkages (for example, between local and national institutions) and the ability of institutions to incorporate new knowledge and innovative behavior, without which institutions may fail to serve their functions and dissolve.[7] Second, the fishing port at Belford has been the object of earlier studies, providing context for analyzing the texts of the interviews.[8] Finally, the extensive damage experienced at the Belford cooperative provides an opportunity to observe the process of resilience or "bouncing back."

BELFORD SEAFOOD COOPERATIVE, BEFORE SANDY

Co-op fishermen at Belford self-identified as gillnetters, trawlers, and crabbers, and they caught fluke (summer flounder), scallops, menhaden, whiting, porgies (scup), skates, dogfish, lobsters, crab, and squid. There are about thirty members of the Belford Seafood Cooperative, who jointly own the co-op. Other people crew the boats or work for the cooperative but do not have a stake in its ownership. Among the thirty members, only about half are active, meaning that they land their fish at the co-op's dock, where the fish are either packed into boxes with ice and sent to regional wholesale markets, or sold whole at the co-op's own small retail market. The pre-storm infrastructure of the Belford Seafood Cooperative consisted of one main building, an ice machine building, and a refrigerated storage shed. The main building housed the co-op's office, the fish market, a large packing shed, and a restaurant with a large deck facing the inlet. Some commercial fishermen keep their boats at the dock nearest the co-op building. Other members of the co-op, as well as fishermen who sell their catch to another business at the port, keep their boats at a municipal dock or at their private docks on both sides of the inlet, where Compton's Creek connects with Raritan Bay, part of the New York Harbor Complex. The fishermen we interviewed typically worked on or owned numerous boats before their current one.

The Belford Cooperative may have been particularly vulnerable to Sandy. First, its location, on the southern side of the vast complex of bays between New Jersey and New York, placed it in the area of the highest storm surge on October 29, 2012. Second, its financial situation was tenuous. Since the late 1970s, the commercial and recreational marine fisheries of New Jersey have gone through a major transition from a mostly unregulated, open-access system to a tightly regulated one, where the cumulative effects of the process of "creeping enclosure" (limiting fishing rights to a set of approved fishermen) have made fishing increasingly difficult.[9] The port of Belford, which is located inside a polluted bay and several miles away from the open ocean, and has small-scale fishing enterprises that have not been able to engage in long-distance fisheries nor invest in the more lucrative limited-entry fisheries, has been hit particularly hard in this process. It has been struggling financially for many decades, and because of that, the co-op was hard-pressed to muster the capital needed to quickly recover from the storm.[10] One might say that the Belford Cooperative was not likely to be economically resilient, but as we shall discuss, cultural resilience, expressed in how people talked about themselves and the co-op (their social bonds), helped to sustain the recovery process.

IMPACTS OF HURRICANE SANDY

Just before Sandy hit, the commercial fishermen of Belford perceived themselves as ready for the storm. Two co-op fishermen remarked that they prepared for Sandy the same way they prepare for any nor'easter or intense storm coming from the northeast. Major weather events are par for the course for commercial fishermen. A co-op fisherman said, "I knew the tide in here [the inlet] was going to be astronomical, and it was." Each crew tied down its own boat while the dock employees prepared the building where the office, fish market, packing shed, and ice-making equipment are located. Unfortunately, as the manager explained on November 15, 2012, they had not taken enough precautions in the office, just moving computers and files to the tops of file cabinets without anticipating that the rising water might upend the file cabinets and destroy the computers and papers. Little was or could be done for the extensive deck used for a restaurant.

Despite these preparations, the port of Belford was devastated by the storm. Belford experienced profound damage at the co-op, including losses to the electrical system, ice machines, and docks. The restaurant, not owned by the co-op but on land owned by it, was totally destroyed. Boats and gear were damaged. Directly after the storm, fishermen also faced problems with low market prices as well as lost time for fishing.

Damage to Fishing Vessels

Numerous fishermen at the Belford Cooperative mentioned with pride that the co-op did not lose a single boat in the storm. Because their fishing vessels are relatively large for the East Coast (from thirty-five to more than seventy feet in length), they were not hauled out of the water beforehand (figure 6.1). Some added, after describing how they tied up their boats before the storm struck, "That's all you can do." The surge was extremely high at Belford (as elsewhere in New York Harbor), thought to have been higher than eight feet in Belford's inlet, but it was not enough to throw the boats up out of the water—a miracle some thought (figure 6.2).

FIGURE 6.1. Boats tied up at dock.

FIGURE 6.2. Fisherman explaining just how narrow the escape was.

Damage to the Port at Belford

Interviewees described the damage at Belford as catastrophic, as did media coverage.[11] Eight-foot waves came into the inlet, and water rose to above five feet in the packing shed and office, which are already several feet above high-tide mark. In addition to the loss of the restaurant and significant damage to the dock and main office building (figures 6.3 and 6.4), fishermen reported losing personal gear, such as nets, generators, and totes in which to store fish while at sea. The computers and files in the co-op's office were also destroyed.

Co-op members spent considerable time and effort explaining the damage at Belford, using illustrative descriptions. One fisherman said the deck for the restaurant moved the distance of "about five football

FIGURE 6.3. Cooperative member marking the level to which the storm surge rose along interior of the packing facility.

fields. . . . It looked like a roller coaster" because it was so bent out of shape (figure 6.5). Another member said that the best way to describe the damage was as if "a jetliner crashed in our [the co-op's] parking lot." Another man said the damage at the co-op was the "biggest mess I've ever seen" (figure 6.6). When asked what the co-op looked like on his first trip back after the storm, another fisherman said it was "like you take everything you own, put it in a trash can, dump it out and pour water and oil all over it. That's the only way I can describe it."

In addition to the physical destruction at the co-op, commercial fishermen in Belford also reported Sandy-induced damage to the inlet used to sail in and out of the port. They said that the inlet was in need of dredging prior to Sandy and estimated that the storm brought in 25 percent more silt,

FIGURE 6.4. Chairs left behind where the restaurant used to be.

greatly restricting when they could go out to sea and return to port, with some needing at least half a high tide to safely navigate the channel. There was a desperate need expressed by some fishermen for aid to dredge these waterways. "We're not going to be operating unless it is dredged." Not only did this affect the ability of fishermen who dock at that port to enter and exit; there was also concern that other fishermen would be discouraged from unpacking at that port once word got out that their channel was dangerous, impacting an important revenue stream for the co-op. The nearby port of Point Pleasant Beach (thirty-five miles south, by the mouth of the Manasquan River) had similar issues and concerns.

Because of Sandy, many fishermen at the Belford Cooperative explained that they lost time fishing, owing to a variety of factors. Co-op members spent time immediately after the storm cleaning up debris and making

FIGURE 6.5. Restaurant dock in shambles, two weeks after the storm.

repairs to their personal fishing property. One member reported significant damage to his personal dock. Repairing the dock was his first priority after the storm because he was unable to get to his boat otherwise. No fisherman interviewed lost a vessel, but some vessels sustained damages. Beyond repairs to personal property, many fishermen were not able to fish until at least the basic physical infrastructure of the co-op was repaired. For example, commercial fishermen require large volumes of ice to keep their catch fresh both while at sea and once they have returned to port to ship the product to market. Generating this volume of ice for the co-op requires ice machines, coolers, and a functioning electrical system—all of which were severely damaged by Sandy. Finally, those who experienced damage to their homes prioritized those repairs over repairs required to return to fishing. For one co-op member whose home was badly damaged, this meant

FIGURE 6.6. Cooperative Manager Joseph Branin in storm-damaged office.

losing several months of fishing. Fishermen reported being out of work from ten days to four months. As a result, many of them missed out on one of the critical fishing seasons of the year, for fluke.

Damage to the Fishing Grounds

Media and agencies have focused on damages to coastal buildings and infrastructure, but our research has found considerable concern among fishermen about effects of the storm on ocean conditions and fish availability. For those who went fishing shortly after the storm, "the stocks were not there." They attributed changes to the bottom and fish migration patterns to Sandy, generally feeling that fluke was impacted the most by these changes. When asked if they were catching quotas, one responded "sometimes, depending on the fishery." The gillnetters seemed to have a harder time catching quota than other gear types did. They mentioned getting debris in gillnets after the storm and described it as "an incredible mess." There was also no bait fishing in the spring. Some were able to catch their quotas farther offshore but burned more fuel, which affected their revenues. Fishermen interviewed in September 2013 were still seeing impacts from Sandy. The "patterns of fishing have been disrupted. . . . Fish have to get used to what's new. Then fishermen have to get used to that." When asked how fishing is now, one fisherman responded, "Still off. Different. It's different. Fish . . . it's like they have gotten off course." Fishermen struggled throughout the summer with what was described as "poor fishing." Income lost in the weeks after the storm continues to put a burden on the fishermen, as one mentioned losing about $50,000, equal to a year's income.

NARRATIVES FROM BELFORD: I AND WE

Two distinct narratives about resilience emerged during our interviews with fishermen at the Belford Seafood Cooperative. One was a personal narrative about fishing as part of a person's identity, and the other was a narrative about the history of the co-op and its ability to survive. We call the latter a community narrative. Both of these narratives are reflective of cultural identities, as fishermen and as co-op members. We did not prompt people to talk about their personal and group identities; rather, we found that language about personal and group identities emerged on its own. These two narratives, personal and community, although present in other fishing ports studied, did not combine as strongly elsewhere. The consistency of this linking of personal and group (co-op) identities in the interviews at Belford's cooperative is meaningful, and we argue that it supported cultural resiliency in the aftermath of Sandy.[12]

Fishing as Personal Identity

The first narrative from Belford was about fishing being part of individual fishermen's identities. Interviewees often talked about the genesis and the permanence of their fishing lives. First, they described how they became fishermen, with many coming from fishing families. Fishermen often named their boats after a close family member, such as a wife, niece, or aunt. One fisherman named his boat after his niece who was born a month before the boat began fishing; he remembers how old the boat is because of his niece's birthday parties. Sometimes members mentioned the age at which they started fishing, the length of time they have spent fishing, or the specific year in which they started fishing. Even those who did not come from a fishing family strongly tied their identities to being a fisherman. For example, one co-op member mentioned that despite not being from a fishing family, he got into the business because he grew up a half mile from the docks and started unloading the boats and working in the ice house when he was just fifteen. Another fisherman said he started working at a fish processing plant when he was so young that he had to use his brother's name to conceal his age. Respondents from other ports in New Jersey and New York less frequently talked about being from a fishing family.

In addition to describing how they became fishermen, interviewees from Belford also described the permanence of fishing in their lives. Some noted that they had been fishing their whole lives or that they were lifelong fishermen. Some said fishing is a "way of life" or that there is no alternative career choice for them. Others used their identity as fishermen to explain their behavior after Sandy. For example, when asked when he first came down to the co-op after the storm, one fisherman said, "11 that night. I'm a fisherman." His response suggests that because he is a fisherman, it should be obvious that he would immediately check on his boat after the storm.

Previous research on fishing communities indicates that characterizing fishing as a way of life is a common theme throughout the industry and is not unique to Belford.[13] However, respondents from the Belford Cooperative consistently drew on this narrative when describing their experiences of Sandy in a way that was not observed in other ports. We observed another narrative from some interviewees in other ports who speak of fishing as a business. A commercial fisherman from Point Pleasant Beach said, "We are no different than Jenkinson's [a nearby tourist destination]. It's a business. A boat is a floating store." We did not hear any fishing as a business comments in Belford. Given the extreme physical damage sustained at Belford in comparison to less significant damage elsewhere, it is difficult to draw clear causal relations. If the damage at the Point Pleasant Beach Cooperative, for example, had been greater, a similar theme may have emerged there. In the absence of comparable damage, a direct comparison cannot be made. Nonetheless, the overall message from Belford was that fishing is a cemented part of people's identities, not easily altered by the impacts of Sandy, despite the severe damage caused by the storm.

History and Persistence

A second narrative that emerged from interviews at Belford focuses on the long-term resilience of the fishing port and the cooperative, and therefore the ability to persist in spite of the extensive damage caused by Sandy. First, co-op fishermen commonly used the term "we" to refer to the entire cooperative. This sense of unity came through in many of the interviews, with comments such as, "Everyone at the co-op is looking out for everyone else. That's how we get by." We believe that the social organization at the Belford Cooperative may help explain this very broad meaning of "we." There is little diversity of roles among Belford Cooperative fishermen, with most interviewees being both boat owners and boat captains. Also, there are many family connections between members at Belford. Five of the ten commercial fishermen and staff interviewed who were associated with the co-op had at least one family member (brother, father, or son) who belonged to or worked at the co-op. Family can be an important source of social capital;[14] thus, together, the homogeneity of roles and family connections support strong ties between members of the co-op.

This use of "we" to refer to the cooperative at Belford is distinct from how the pronoun was used by respondents in other ports. Significantly, interviewees at the other cooperative in New Jersey, Point Pleasant Beach, did not use "we" to talk about the co-op as a group, but rather to discuss the crews of their individual fishing vessels. It is uncertain if this phenomenon is also related to damage severity. Had there been more damage at Point Pleasant Beach, requiring greater collective action, there might have been a strong discussion of "we, the co-op."

The second way in which the identity of the cooperative emerged was in the way fishermen described their fellow co-op members. A fisherman of forty years mentioned that despite the fact that the cooperative lost the restaurant, a major source of revenue, and that they won't be able to bring that back, "we're [co-op] a resilient bunch." He went on to add that because they've been there "for generations," they would recover.

Although frustration with the lack of assistance from the federal government was a theme we noted across all sectors and all ports, at the Belford

Cooperative it seemed to uniquely strengthen the resolve to recover from Sandy, solidifying the concept of "us" as the co-op and everyone else, including the federal government, as outside of that "us." Belford Cooperative members looked to the federal government and private insurance companies to provide recovery assistance, but the Belford interviewees, like all fishermen interviewed for this project, expressed a great deal of frustration in dealing with these organizations. Many commented that all the Federal Emergency Management Administration (FEMA) was offering were loans with interest rates higher than what they could get through their local bank, which is where they ended up taking out a loan. Further, despite negative feelings about lack of assistance from FEMA and the strain felt from fishing regulations, interviewees often provided accounts of how they could overcome these obstacles through hard work. One Belford Cooperative member stated, "We'll get there [recovery] no thanks to the government." Another co-op fisherman predicted, "We'll recover. We always have. We are a bunch of hardworking people. No one has ever helped us before."

Belford Cooperative members also drew on the past when speculating on the future of the co-op. Specifically, members highlighted the fact that the co-op had experienced and survived many other disasters. In addition to dealing with storms and iced-in conditions, the port of Belford has a long history of community identity grounded in decades of resistance to fishing regulations that made commercial fishing difficult in the waters close to the port.[15] They have to compete with sport fishing, shipping, marine pollution, increasingly stringent fishery management regimes, and other challenges to trying to fish for a living at a small scale—that is, without investments in large, industrial-scale fishing operations—in one of the world's busiest and most urbanized regions. In the mid-1980s, the fishing port of Belford, led by the co-op, engaged in a successful struggle to ward off the sale of much of the waterfront and land by the owner of a former fish processing plant; that sale would have transformed the port into a pleasure boat harbor and seaside condominiums. Instead, the co-op was able to get assistance from the Port Authority of New York and New Jersey to gain secure title to the land and waterfront needed for the fishery and help with bulkheading, as well as zoning protection. Nonetheless, the co-op must continue struggling to survive as a not always attractive business in a neighborhood increasingly marked by upscale housing and well-off commuters, using nearby ferry service to Manhattan.[16] Fishermen and fishing families of the port of Belford are also the focus of many planning efforts, as yet unsuccessful, to obtain assistance in maintaining and developing the fishing port and community.[17]

Recovery from Sandy is just one in a long history of recoveries from other calamities and threats. One young fisherman in the co-op said that many Belford fishermen "had been through this [extensive damage] before. . . . The guys have survived there for one hundred years. I don't see why this would stop them. . . . It's like having a bad day, you get over it." Another fisherman commented on how the fishing port has been there for one hundred years and how the long history of the fisheries in Belford would provide the necessary fodder for recovery. In spite of the fact that there was "no joy in the air" the day after Sandy, everyone knew the co-op would recover. He added, "We'll get back together. We have for one hundred years." Interestingly, in talking about "we," members of the co-op often conflated the history of the co-op, which was created in 1953, with the much longer history of the port. This conflation, rather than being a point of contention, was used as a point of pride and ultimately a source of cultural resilience.

CONCLUSION

The ability to cope with and recover from this complex set of events and threats, highlighted by our discussion of Belford's co-op fishermen, is based on adaptation and resiliency, albeit with a touch of fatalism. Despite the significant losses many fishermen incurred, some maintained their resiliency. For example, one fisherman who sustained a heavy loss stated, "This is the price of doing business, we're fishermen, we have to live on the water here. . . . I don't know what else we can do." Fishermen interviewed in September 2013 were still experiencing impacts from Hurricane Sandy. The theme of adaptation and resilience appeared then too, in an interview: The "patterns of fishing have been disrupted. . . . Fish have to get used to what's new. Then fishermen have to get used to that."

The identity themes observed at the Belford Cooperative were seen in interviews at other ports, but not with the consistency and cohesion that came through at Belford. In particular, the idea of fishing as a way of life was expressed by fishermen at other ports, but as part of a description of their personal histories. What was unique about Belford was that the co-op members took that personal identity and used it as a foundation for building an identity as a community through the institution of the cooperative.

This strong sense of community at Belford preceded the storm; it was not created by Sandy. The damage done by the storm and subsequent recovery efforts did not transform a group of independent fishermen into a community working together. Rather, Sandy and its aftermath reinforced an existing narrative about what it means to be a member of the Belford Cooperative, to demonstrate to themselves, once again, that they, as a community, have the will and ability to "bounce back," strengthening their image of themselves as a permanent fixture in the landscape of fishing in the northeastern United States. Their personal identities as fishermen and their identity as a fishing community are thus sources of cultural resilience. Whether that cultural resilience is sufficient for economic recovery is uncertain and can only be assessed in the future. It is possible that an event could simply exceed the capacity of an institution like the co-op to recover. Sandy did not overcome the co-op, although its effects on individual fishermen, combined with other hardships, could permanently harm some individuals.

Therefore, the storm was not a socioculturally transformative event for the fishing community at Belford. Rather it reinforced the existing sense of community that focused on the cooperative and facilitated rebuilding toward the previous status quo.[18] The co-op members at Belford already self-identify strongly as fishermen, and storms are a normal part of that identity. Fishermen expect storms. Despite their obvious and direct reliance on the ocean, it is unclear if any natural process associated with the ocean could meaningfully impact their individual or community identities. Rather, damage from the storm may have reinforced and strengthened these identities. The extent to which Sandy may have been an economically transformative event, leading to altered development patterns remains uncertain. However, the co-op's history of fighting development and gentrification suggests the port of Belford may survive economic transformation as well, at least for now.

LESSONS

This case yields two lessons about storm recovery that can affect the vitality and culture of coastal communities engaged in commercial fisheries:

- Understand the local institutional structure of fishing communities (that is, is there a cooperative or similar fishermen's group?).

As this study shows, fishing communities can initiate the recovery process after a storm, particularly in the earliest days. But for outside agencies and organizations that eventually participate in this process, working with fishermen poses different challenges than working with other businesses. Fishermen may fish on their own every day, but they depend on other businesses, organizations, and institutions for equipment, information, and selling their catch.

- Work closely with existing local institutions, including cooperatives, in identifying problems and coming up with plans for restoration.

A commercial fishery can be supported best through its key institutions, such as cooperatives, which take on new roles during and after an emergency. Practices of mutual aid differ across fisheries and communities, and officials can enhance these practices by working with existing institutions to speed recovery.

NOTES

We are grateful to our interviewees for sharing valuable insights, our colleagues Angela Silva and Ariele Baker for interview and analysis support, and the National Science Foundation "RAPID" program (BCS 1318074) for financial support. Interviews for this research were conducted under conditions reviewed and approved by the institutional review boards of our home institutions. Interviewees granted us their informed consent to quote or paraphrase their comments.

1. Chris Christie, letter to the Honorable Rebecca Blank, Acting Secretary of Commerce and Deputy Secretary of Commerce, November 13, 2012, accessed December 5, 2012, http://www.nj.gov/dep/docs/fishery_letter.pdf.
2. Julie Urquhart and Tim Acott, "A Sense of Place in Cultural Ecosystem Services: The Case of Cornish Fishing Communities," *Society and Natural Resources* 27 (2004): 3–19;

Kai M. A. Chan, Terre Satterfield, and Joshua Goldstein, "Rethinking Ecosystem Services to Better Address and Navigate Cultural Values," *Ecological Economics* 74 (2012): 8–18.

3. Todd A. Crane, "Of Models and Meanings: Cultural Resilience in Social-Ecological Systems," *Ecology and Society* 15, no. 19 (2010): http://www.ecologyandsociety.org/vol15/iss4/art19. Please note that throughout this chapter we will refer to *fishermen* rather than *fishers*, as most U.S. men and women who fish commercially prefer this designation. A *fisher*, they note, is actually a marten (*Martes pennanti*), a member of the weasel family (Mustelidae); therefore, they find this term derogatory.

4. Susan L. Cutter, Lindsey Barnes, Melissa Berry, Christopher Burton, Elijah Evans, Eric Tate, and Jennifer Webb, "A Place-Based Model for Understanding Community Resilience to Natural Disasters," *Global Environmental Change* 18 (2008): 598–606.

5. Emily Chamlee-Wright and Virgil Henry Storr, "Social Capital as Collective Narratives and Post-disaster Community Recovery," *Sociological Review* 59, no. 2 (2011): 266–282; Daniel P. Aldrich and Yasuyuki Sawada, "The Physical and Social Determinants of Mortality in the 3.11 Tsunami," *Social Science & Medicine* 124 (2015): 66–75, http://www.purdue.edu/discoverypark/climate/assets/pdfs/Daniel%20Aldrich.pdf; AP-NORC, *Two Years after Superstorm Sandy: Resilience in Twelve Neighborhoods*, Associated Press–NORC Center for Public Affairs Research (NORC at the University of Chicago, 2014), http://www.apnorc.org/PDFs/Sandy/Sandy%20Phase%202%20Report_Final.pdf; Trevor Thompson, Jennifer Benz, Jennifer Agiesta, Kate Cagney, and Michael Meit, "Resilience in the Wake of Superstorm Sandy," Associated Press–NORC Center for Public Affairs Research (NORC at the University of Chicago, 2013); Daniel P. Aldrich, "The Power of People: Social Capital's Role in Recovery from the 1995 Kobe Earthquake," *Natural Hazards* 56, no. 3 (2011): 595–611, http://link.springer.com/article/10.1007%2Fs11069-010-9577-7.

6. Elkeand Herrfahrdt-Pähle and Claudi Pahl-Wostl, "Continuity and Change in Social-Ecological Systems: The Role of Institutional Resilience," *Ecology and Society* 17, no. 8 (2012): http://www.ecologyandsociety.org/vol17/iss2/art8.

7. Marie-Caroline Badjeck, Jaime Mendo, Matthias Wolff, and Hellmuth Lange, "Climate Variability and the Peruvian Scallop Fishery: The Role of Formal Institutions in Resilience Building," *Climatic Change* 94 (2009): 211–232.

8. Bonnie J. McCay, "Optimal Foragers or Political Actors: Ecological Analyses of a New Jersey Fishery," *American Ethnologist* 8 (1981): 356–382; McCay, "The Pirates of Piscary: Ethnohistory of Illegal Fishing in New Jersey," *Ethnohistory* 31 (1984): 17–37; Bonnie J. McCay and Marie Cieri, *Fishing Ports of the Mid-Atlantic: Report to the Mid-Atlantic Fisheries Management Council, Dover, Delaware, April, 2000* (New Brunswick, NJ: Department of Human Ecology, Cook College, Rutgers, 2000); Bonnie J. McCay and Debra Mans, "Coastal Gentrification: The Belford Case," presentation at the Northeast Region Coastal Managers Conference, Long Branch, NJ, February 19, 2004.

9. Grant Murray, Teresa Johnson, Bonnie J. McCay, Satsuki Takahashi, and Kevin St. Martin, "Cumulative Effects, Creeping Enclosure, and the Marine Commons of New Jersey," *International Journal of the Commons* 4 (2010): 367–389.

10. Joseph Branin, manager of the Belford Seafood Cooperative, personal communication to Angela Oberg and Julia Flagg, November 15, 2012.

11. Mark Di Ionno, "Generations of Fishermen Stuck on the Shore after Sandy Walloped Their Business," *NJ.com Blog*, November 29, 2012.

12. Chamlee-Wright and Storr, "Social Capital as Collective Narratives and Post-disaster Community Recovery."

13. Richard B. Pollnac and John J. Poggie, "The Structure of Job Satisfaction among New England Fishermen and Its Application to Fisheries Management Policy," *American Anthropologist* 90 (1988): 888–901; Helen Mederer, "Surviving the Demise of a Way of Life: Stress and Resilience in Northeastern Fishing Families," in *The Dynamics of Resilient Families: Resilience in Families*, ed. Elizabeth A. McCubbin, Annie I. Thompson, and Jo A. Futrell (Thousand Oaks, CA: Sage, 1999), 203–235; Ryan Kelty and Ruth Kelty, "Human Dimensions of a Fishery at a Crossroads: Resource Valuation, Identity, and Way of Life in a Seasonal Fishing Community," *Society and Natural Resources* 24 (2011): 334–348; Ruth Williams, "The Socio-Cultural Impact of Industry Restructuring: Fishing Identities in Northeast Scotland," in *Social Issues in Sustainable Fisheries Management*, ed. Julie Urquhart, Tim G. Acott, David Symes, and Minghua Zhao (Amsterdam: MARE Publication Series, no. 9, 2014), 301–317.

14. Margaret M. Bubolza, "Family as Source, User, and Builder of Social Capital," *Journal of Socio-Economics* 30 (2001): 129–131.

15. McCay, "Optimal Foragers or Political Actors"; McCay, "The Pirates of Piscary."

16. Judy Peet, "Plans for a Ferry Service and Townhouses Threaten the Way of Life in the Small Fishing Village of Belford," *Star-Ledger* (Newark, NJ), February 11, 2001.

17. Louis Berger Group, *Port of Belford Economic Feasibility Study and Conceptual Development Plan–Township of Middletown, Middletown, New Jersey*, 2009, http://www.middletownnj.org/DocumentCenter/Home/View/103; Barbara Jones, "Belford: A Mid-Atlantic Fishing Community Facing Change," *Practicing Anthropology* 25 (2003): 14–18; Jones, "Teetering towards Economic Sustainability: Alternatives for Commercial Fishermen," *Journal of Ecological Anthropology* 13, no. 1 (2009): 73–77; Edward J. Bloustein School of Planning and Public Policy, *The Belford Neighborhood Revitalization Plan*, Fall 2012, http://bloustein.rutgers.edu/wp-content/uploads/2012/12/The-Belford-Neighborhood-Revitalization-Plan.pdf

18. Chamlee-Wright and Storr, "Social Capital as Collective Narratives and Post-disaster Community Recovery."

PART III PLANNING FOR CHANGE?

7 • GREEN GENTRIFICATION AND HURRICANE SANDY

The Resilience of the Green Growth Machine around Brooklyn's Gowanus Canal

KENNETH A. GOULD AND TAMMY L. LEWIS

On Friday, October 26, 2012, an "Introduction to Urban Sustainability" class from Brooklyn College participated in a podcast walking tour of the Gowanus Canal. Standing on the banks of the canal, they listened as the recorded voice for their self-guided tour explained what would happen if a hurricane-induced storm surge hit New York Harbor. They were instructed to look up the gradually sloping street from the edge of the canal and imagine what thirteen feet of sewage-laden, toxic floodwater would look like as it moved up the slope, into basements, across the thresholds of first floors of residences, and then up toward the second floors of homes and businesses directly adjacent to the canal. Precisely that scenario would unfold the following Monday, as Hurricane Sandy hit at high tide. Some of those students would lose their homes to flooding in the Rockaways. When their college reopened more than a week later, seemingly abstract theoretical scenarios of urban vulnerability and resilience gave way to relief efforts and grounded discussions of how the city would move forward. The podcast also needed revision, as the warnings about worst-case scenarios had to be replaced by the recounting of the actual disaster, including flooding, fires, and the loss of key transportation infrastructure.

In 2012, the fate of the ecological health of the Gowanus Canal was at a crossroads. Hurricane Sandy was only the latest blow to a site that had been treated like a dump for decades. However, like other waterfront areas, even contaminated ones, the pressure to develop real estate had finally forced the city and developers to consider cleaning it up for development. In 2010, the canal gained federal Superfund status, though little had begun by the time Sandy struck. The ecological restoration of the Gowanus Canal could provide greater opportunities for real estate developers to invest in the area and increase housing values. But given the devastating nature of the Sandy-induced major flood event, would they continue with the plan?

In the end, real estate interests proposing projects they claim are green, leading to green gentrification, are stronger than the storm. This was true of development around the Gowanus Canal and other developments in flood zones around New York City, such as Hudson Yards. The irony is that this green growth machine, while ostensibly creating more environmentally sustainable urban spaces, is socially unsustainable. Current low- and middle-income residents will be replaced by higher-income residents drawn to the revitalized, greened, environmental resource, thus leading to what we call green gentrification. It is also ecologically unsustainable. It leads to increased development and settlement in flood-prone areas, a trend made more worrisome by the growing threat of climate change. Although this case study discusses some features of development politics that may be distinctive to New York City, it also presents concepts that can be used to analyze how flood zone development is intensifying in nearly every major waterfront city around the world.

TERMS AND LITERATURE

We use the term *green gentrification* to describe processes facilitated by the creation or restoration of environmental amenities that draw in a wealthier group of residents. There are many causes of gentrification, including historical architecture, proximity to transportation, and cycles of investment.[1] Green-led redevelopment is often intentional, as investors and public officials create new or renewed green spaces, including parks, as a means to raise property values.[2] A definition of gentrification we find useful for its emphasis on distributional impacts is by Tom Angotti in his book about New York City real estate processes, *New York for Sale: "As* tenants and small business owners invest their time and money to gradually upgrade their neighborhoods, real estate investors become attracted to these areas and anxious to capitalize on the improvements. As investors large and small move in, they effectively appropriate the value generated by others. This is the essence of what is now known as gentrification. It is not simply a change in demographics. It is the appropriation of economic value by one class from another."[3]

"Green" gentrification is not necessarily due to the actions of "tenants and small business owners" themselves, as it often comes primarily from outside public and private investors who appropriate the value of an unrevitalized environmental resource. In this sense, it is the appropriation of the economic value of an environmental resource by one class from another, a process that has happened throughout New York City.[4] The concept of green gentrification also grows out of literature on environmental injustice showing that environmental negatives in society, such as toxic pollutants and locally unwanted land uses, are disproportionately found in minority and poor neighborhoods.[5] The existence and structure of environmental injustice in the United States rests on the fact that residential neighborhoods, including those in Brooklyn, are segregated by race and class.[6]

Race and class residential segregation set the stage for unequal distribution of urban environmental positives and negatives.[7] The structure of housing markets functions to reinforce class-based distributions of environmental benefits and detriments. If an area believed to be free of environmental hazards is found to be contaminated, those with greater wealth will be able to move to a less hazardous location.[8] Conversely, if environmental amenities are created or restored in a community, the value of that real estate increases and housing prices are pushed upward. Residents unable to increase the share of their incomes spent on rent are pushed out of their neighborhoods, increasing the gap between the socioenvironmental haves and have-nots.[9]

The social movements that have developed to address environmental injustices, and the academic literature that is in dialogue with these movements, have increasingly moved toward what Julian Agyeman calls a "just sustainability" paradigm.[10] This shift refocuses research questions toward who gets the environmental goods, such as parks, clean air and water, and access to waterfront resources.[11] The Gowanus Canal is especially interesting because it has been considered an environmental bad and is being recast and restored to become an environmental good. This shift is being accompanied by shifts in the adjoining neighborhood demographics.

There are few published accounts examining how neighborhood constituencies change due to urban ecological restoration.[12] The creation of community gardens has been shown to increase neighborhood property values,[13] although it is not clear if constituencies change. According to Jonathan D. Essoka, constituencies do change after the redevelopment of brownfields, sites where contamination has been removed or sequestered in some fashion. His analysis of demographic changes at sixty-one brownfield redevelopment sites concludes: "Statistical techniques revealed that gentrification is often a consequence of brownfields redevelopment. The data demonstrates that for Blacks and Latinos, while their overall metropolitan populations increase or stay the same, local brownfields revitalization forces their [local] number to decrease."[14]

Thus, due to the intersections of housing segregation, processes reproducing environmental injustice, and the operation of markets and actors that form what Harvey Molotch and John R. Logan characterize as the urban growth machine,[15] the creation or restoration of an environmental good will increase environmental inequality in the absence of policy intervention to do otherwise. We call this a green growth machine in which ecological restoration forms the basis of the local growth coalition to initiate site-specific green urban redevelopment. In the drive to extract increased value from real estate, elites harness environmental concerns to generate publicly funded environmental amenities and restoration.[16] Those environmental improvements raise the value of real estate investments, enable investors to quickly resell properties at a profit, and thus promote gentrification, increasing capital gains to private developers and tax revenues for the state. The green growth machine is thus part of an urban redevelopment treadmill in which neighborhoods are destroyed by sustainability initiatives and recreated in ways that benefit nonresidents.

BROOKLYN'S GOWANUS CANAL

The Gowanus Canal's historic transformation from tidal marshland, to toxic dump, to urban environmental amenity illustrates the dynamic relationship between environment, economy, and social distribution. Starting with the extractive development of oyster harvesting, to milling, to industrialization, and now residential real estate development, the primary engine of ecological and neighborhood change has been the interests of the growth coalition of each era, and each era has required or produced a different neighborhood composition. Despite Sandy's momentary disruption of the redevelopment process, elites push on with development plans. A brief history provides some context for the various ways that the site has been contaminated. In light of this long history as a dumping ground, it is all the more remarkable that the green growth machine is pressing forward with a clean-up for luxury real estate development.

A Brief History of Gowanus

Ostensibly named for the leader of the Canarsee band of the Lenni Lenape Indians, Gowanus Bay was a tidal inlet of saltwater marshland. By the mid-1800s, Brooklyn was the fastest-growing city in the United States. Local elites argued for the construction of a canal to drain marshlands, raise property values, and serve local industry. The decision to create a 1.8-mile inland waterway with no flow-through would have dramatic consequences for water quality, public health, and the area's long-term development.

The state-directed construction of the Gowanus Canal, completed in 1869, created one of the nation's first planned industrial development districts. Industries quickly arose and expanded, including coal gas manufacturing plants; cement works; paper mills; paint, ink, and chemical plants; and sulfur producers. By 1880, Brooklyn was the fourth-largest industrial city in the country, attracting a large working-class immigrant population that was then subjected to the toxic mélange of airborne effluent and routine flooding.[17]

With growth of the surrounding neighborhoods at a rate of nearly seven hundred new buildings a year by the late nineteenth century, population growth quickly outstripped sanitation planning. The raw sewage of surrounding neighborhoods drained downhill to the Gowanus. When Brooklyn constructed the first municipal sewer system in the United States, it expanded the range of sewage catchment and drained human waste directly from affluent uphill brownstone neighborhoods to the open sewer in the working-class Gowanus neighborhood.[18] Calls for some form of remediation resulted in a publicly funded underground "flushing tunnel" that connected the head of the canal to New York Harbor, which became operational in 1911. Flushing the canal with seawater reduced the stench and the accumulation of both raw sewage and industrial effluent, although the deposition of human waste into the canal necessitated routine dredging to keep the channel open. The peak of the Gowanus industrial boom occurred around the end of World War I, when the Gowanus was the nation's busiest and most polluted industrial canal.

Following World War II, the scale and methods of commercial shipping shifted, as containerization required larger ships, deeper harbors, more expansive yards, and an increasing reliance on trucking to and from ports. Shipping shifted from Brooklyn to the expanding port facilities of New Jersey. Brooklyn's waterfronts would become underutilized and fall into decay for the next half century. In 1955, the Army Corps of Engineers suspended regular dredging of the canal.

In the early 1960s, the flushing tunnel broke down. With the canal no longer vital to the city's economy, and its formerly European population being replaced by largely Hispanic immigrant groups, the city proved financially and politically uninterested in repairing the canal's cleansing mechanism, especially as the city headed into the fiscal crisis of the 1970s. Raw sewage would now remain in the stagnant water to mix with a legacy of industrial toxins and the spillage from the diminished oil industries that remained active along the canal.

The city's abandonment of the canal cannot be separated from the change in the neighborhood's demographics, as both cause and consequence. In 1948, Robert Moses's Slum Clearance Authority had initiated the construction of the Gowanus Houses project, and in 1966 the Wyckoff Gardens housing project was built. Real estate investment had shifted to the suburbs. The Gowanus was now a place where sewage and low-income populations were socially invisible to the city's elites.

But some long-term residents refused to walk away. The Gowanus Canal Community Development Corporation (GCCDC) led the call for redevelopment, in an early vision of this industrial and municipal "end pipe" as an environmental amenity. Although this vision of the Gowanus as "the Venice of Brooklyn" seemed quixotic, some developments in the 1970s lent hope.

The Federal Water Pollution Control Amendments of 1972, and the Clean Water Act of 1977 provided the legal basis for local demands to clean up the canal (table 7.1). In 1975, some dredging of the canal resumed. However, oil spills and fires from the remaining industries and ongoing sewer discharge, combined with the city's budget crisis, led to little water quality improvement or redevelopment. Under federal pressure, the city constructed the Red Hook Water Pollution Control Plant in 1987 to treat sewage headed for the canal, with mixed success. While the routine dumping of raw sewage into the canal was terminated, fourteen combined sewage overflow outlets (CSOs) linked to the canal continued to dump untreated human waste at every major rain event. The CSOs remain a source of Gowanus Canal contamination and an obstacle to any repurposing of the canal as an environmental amenity.

By the 1990s, the gentrification of brownstone Brooklyn had picked up steam, and rising housing costs up-slope in surrounding Park Slope, Cobble Hill, Boerum Hill, and Carroll Gardens led artists and would-be gentrifying in-migrants to seek out the cheaper housing stock in Gowanus. Out of its fiscal crisis now, the city started to look at the potential redevelopment of Gowanus that the GCCDC and other groups had been promoting. Reconstruction of the flushing tunnel was completed in 1998. Property values in the neighborhood rose roughly 40 percent between 1994 and 1998, capturing the interest of city officials and developers.[19] Local politicians began channeling funds to the GCCDC to produce studies, redevelopment plans,

TABLE 7.1 Selected Federal, State, and Local Policies Affecting Gowanus Cleanup

Federal:	Clean Water Act, 1977
	Comprehensive Environmental Response, Compensation, and Liability Act (CERCLA, aka Superfund), 1980
	President Clinton's Executive Order 12898 addressing environmental justice, 1994
	Environmental Protection Agency's Brownfield Program, 1995
State:	New York Brownfield Cleanup Act, 2003
City:	Brownfield and Community Rehabilitation Act, establishing Office of Environmental Remediation and NYC Brownfield Cleanup Program, 2008

and public open space for the Gowanus (table 7.2). No longer useful to the growth machine as an industrial site, the Gowanus ecosystem was being reimagined as waterfront property that could be reconstructed as an environmental amenity.

The renewed flushing tunnel was now pumping 215 million gallons of seawater through the canal daily, and killifish, crabs, and ducks were appearing in the waterway. In 2002, the U.S. Army Corps of Engineers entered into a cost-sharing agreement with the New York City Department of Environmental Protection on a $5 million ecosystem restoration feasibility study of the Gowanus Canal area. The potential to draw educated gentrifiers had made the ecological restoration of the Gowanus Canal feasible. In 1999, the Gowanus Dredgers Canoe Club began serving as a local environmental monitoring crew. Other local groups also focused on the Gowanus environment and redevelopment. By 2000, Ikea and Lowe's had both expressed interest in locating stores in the area.[20] Lowe's ultimately built there, with the requirement to include a small canal-side public-access park. Between 2003 and 2007, the market price for industrial property in Gowanus rose from $108 to $270 per square foot.[21]

TABLE 7.2 Selected Public Funding for Gowanus Canal and Brownfields in New York City

1999	Assemblywoman Joan Millman provided the GCCDC with $100,000 for a bulkhead study and public access document.
2000	The New York City Department of Parks and Recreation gave the GCCDC $270,000 to create three street-end open space parks as part of the Green Streets program.
2001	Governor Pataki funded a $270,000 revitalization plan.
2002	U.S. Army Corp of Engineers and New York Department of Environmental Protection allocate $5 million for an ecosystem restoration feasibility study.
2002	Congresswoman Nydia Velazquez allocated $225,000 for a comprehensive community development plan for Gowanus.
2002	Borough President Marty Markowitz received $100,000 in state funds to identify areas for habitat restoration.
1996, 2003, 2004, 2005, 2006, 2007	The City of New York has received six grants from the federal government for brownfields assessments, ranging from $200,000 to $800,000 (USEPA, Brownfields and Land Revitalization).

This state and private interest in the Gowanus facilitated expanding populations of crabs and artists. In 2003, the Gowanus Artists Open Studio Tour featured more than seventy resident artists. Artists' events and celebrations of the Gowanus environment became regular features of neighborhood culture. Plans for a "Gowanus Green" housing development on a former coal gasification plant site, a canal-side Whole Foods supermarket, and numerous other residential and retail developments geared toward meeting the needs of new or future Gowanus residents proliferated (table 7.3). The city has been generous in granting zoning changes and promising more changes to facilitate private development plans.

The rush to redevelop Gowanus hit a bump when, in 2010, the U.S. Environmental Protection Agency designated the Gowanus Canal a federal Superfund cleanup site. The cleanup is estimated to cost between $300 million and $500 million and to take up to twelve years to complete. Toll Brothers, a large developer firm, withdrew its residential development proposal, as it had threatened to do if the federal government became involved in the cleanup. Artist colonization, however, remained undeterred, and perhaps the status added to the hip, edgy cachet of the locale. In December 2010, the *Wall Street Journal* ran a story about the area titled, "Superfund Site Morphs into Cultural Hotspot."[22]

TABLE 7.3 Selected Private Proposals for Development around the Gowanus Canal

1998	Brooklyn Commons: 500,000 square-foot sports complex and theater (not developed)
2000	Ikea (developed elsewhere)
2000	Lowe's (developed and opened in 2004)
2003	Gowanus Green, mixed use residential (status unclear, as of 2008 had entered a partnership with the city)
2003	Whole Foods (received support from Community Board in 2011 and completed in 2012)
2006	Toll Brothers (zoning change granted to allow for residential project in 2009; pulled out after Superfund designation in 2010; taken over by Lightstone Group in 2012)
2015	Lightstone Group's 700-unit condominium project is under construction and scheduled to open in early 2016

Hurricane Sandy

Despite bringing relatively little rain, and lower than hurricane speed winds, Hurricane Sandy pushed a thirteen-foot storm surge into New York Harbor, flooded low-lying coastal neighborhoods of New York City, knocked out electrical power in lower Manhattan, flooded a good portion of the subway system, destroyed housing on Staten Island, in the Rockaways, and elsewhere, set an entire neighborhood ablaze, and left thousands of New Yorkers without shelter, heat, power, and functional transportation infrastructure for months, and beyond.[23] In Brooklyn, the storm surge pushed the sea up into and over the low-lying neighborhood of Red Hook and up the Gowanus Canal, sending sewage-laden waters over the banks, up the streets, and into the basements and ground-floor residential, industrial, and commercial spaces of the Gowanus neighborhood. The Third Street and Carroll Street bridges were completely submerged. Floodwater spread out more than a block on either side of the canal, submerging the proposed sites for luxury condo developments. The area between Bond and Nevins Streets was turned into a five-foot-deep lake, inundating homes. Floodwater crashed through art spaces, upending large metal sculptural works and mixing paint and chemicals with untreated sewage. Despite being in Flood Zone A (recently relabeled Zone 1), designated for mandatory evacuation, many Gowanus residents had remained in their homes in an effort to protect their property and their pets. Having been warned to evacuate for Hurricane Irene, which did not produce Gowanus flooding, residents were understandably skeptical of weather hyperbole.

Flooding and power outages took the Gowanus Canal pumping station off-line for thirty-three hours, causing 13 million gallons of untreated sewage to discharge into the floodwaters that still covered the neighborhood. As the flood receded the following day, the stench of polluted water and the sound of generators powering pumps to clear flooded basements filled the neighborhood. City officials sent e-mail messages warning residents to avoid skin contact with contaminated water, but the flood had left most affected residents without Internet access. The Environmental Protection Agency issued the following instructions to residents and volunteers who came to help:

- Remove or pump out standing water.
- Use bleach to kill germs.

- Wear rubber boots, rubber gloves, and goggles.
- Clean hard things with soap and water. Then clean with a mix of 1 cup of household liquid bleach in 5 gallons of water. . . . Scrub rough surfaces with a stiff brush and air dry.
- Wash clothes worn during cleanups in hot water and detergent. These clothes should be washed separately from uncontaminated clothes.

Later test results would indicate that the Gowanus floodwaters contained very high levels of bacteria such as enterococcus, owing to raw sewage discharges. Enterococcus can cause diverticulitis, urinary tract infections, and meningitis. Levels of semivolatile organic compounds of the type known to be a major component of Gowanus Canal sludge (such as polycyclic aromatic hydrocarbons) were found in low levels in some on-land samples and were undetected in others.

Neither enterococcus nor storm surge flooding proved much of a deterrent to the green gentrification process. Little more than a year after Sandy, a sustainability themed Whole Foods celebrated its grand opening on the banks of the Gowanus, and the Lightstone Group had broken ground on a seven hundred–unit apartment complex on the previously flooded development site abandoned by Toll Brothers. What remains to be seen is how much of the working-class employment and population will survive this experiment in urban greening.

The Gowanus "Neighborhood" Today

The borders of the Gowanus neighborhood are defined differently by community organizations, residents, real estate brokers, and census tracts, but to indicate trends in gentrification, we consider real estate practices and data from the census. At this stage in its development, it is common to see properties close to the Gowanus Canal being marketed as part of the already gentrified and well-known areas of Park Slope or Carroll Gardens. The *New York Times* has recently added Gowanus to its searchable neighborhoods in its real estate section, but only one of the large real estate brokers' offices (Corcoran) in Brooklyn has listed Gowanus as a neighborhood: "Gowanus, often referred to as 'G Slope' (because it is just west of Park Slope) is a neighborhood with an industrial, edgy feel, yet Gowanus offers a nice sense of community and a number of homes to buy. . . . The most anticipated real estate development in Gowanus is the area bordering the Canal."[24] The use of the term *G Slope* to refer to this area is unique to Corcoran, a hip new name that capitalizes on Park Slope's appeal and that does not conjure existing negative connotations associated with the canal.

To consider how historical trends have affected the makeup of the Gowanus area, we analyzed the region being most actively pursued by developers. We analyzed census tracts 125, 123, 121, and 117 in terms of demographic shifts and compared them with Brooklyn as a whole for the period from 1990 until 2010 (figure 7.1). While incomes in Brooklyn declined over this period (in real dollars), they rose in all four tracts. Educational levels also rose more dramatically than in Brooklyn as a whole. The cost of housing is perhaps most telling, with both housing prices and rents rising faster than in Brooklyn overall. In sum, areas immediately around the Gowanus are starting to look more like their already gentrified neighbors, Carroll Gardens and Park Slope.

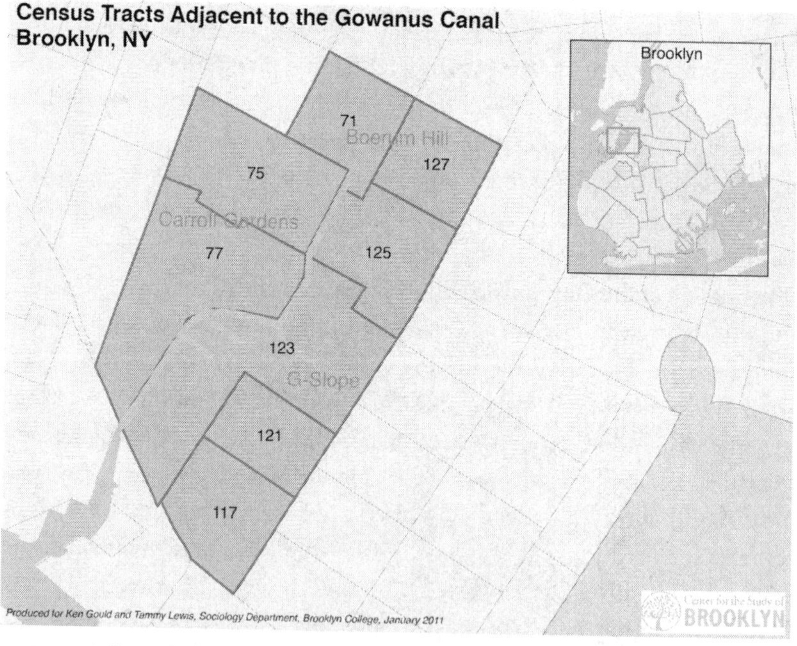

FIGURE 7.1. Census tracts adjacent to the Gowanus Canal. (Source: Center for the Study of Brooklyn.)

GENTRIFICATION AND THE URBAN GREEN GROWTH MACHINE

A confluence of historical conditions and social actors created the context in which the green growth machine would come to see the fetid Gowanus as a wealth-creating opportunity. Deindustrialization had decreased the canal's economic significance as a site of production and employment. An economic boom in Brooklyn, especially after the terror attacks in Manhattan on September 11, 2001, created demand for housing. The neighborhoods around the canal had already gentrified significantly. The canal's neighboring attractions and in-situ potential drew attention from private real estate developers. After a half century, the Gowanus became socially visible to capital again.

Government mandates, initiatives, and rhetoric reinforced these economic trends. For instance, federal water quality legislation (Clean Water Act) required that the Gowanus meet certain environmental standards. Opportunities like New York State's Brownfield Cleanup Program (established in 2003) made funds available to address the contamination of the canal and the lands around it. Two of the ten priorities in Mayor Bloomberg's PlaNYC, his major sustainability initiative, directly affect canal redevelopment. The plan promises to "create more homes for almost a million more New Yorkers, while making housing more affordable and sustainable" and to "clean up all contaminated land in New York."[25] And in 2009, the Obama administration called for a jump-start of a national "green economy," drawing heavily on the advocacy of the Apollo Alliance.

This political-economic context dovetailed with social trends. Upper-middle-class urban residents were supporting sustainability movements, such as urban food movements and seeking housing that developers termed as green. Older community groups, such as the GCCDC, who had long-sought to clean up the Gowanus, found allies in these groups and in the artists who had started to colonize the Gowanus neighborhood. The emerging political opportunity structure coupled with the interest of private developers, the local government, long-term residents, new residents (the artists), and a large potential group of in-migrants converged to make 2010 the moment that the Gowanus Canal's cleanup would begin in earnest and the toxic canal would begin to turn in everyone's imagination into the basis for an urban environmental amenity. This vision was undeterred by the flooding caused by Hurricane Sandy in 2012.

GREEN GENTRIFICATION: STRONGER THAN THE STORM

What are the redistributional impacts of Gowanus remediation and redevelopment? Who will pay for the structural mitigation necessary to keep the residences adjacent to the "Venice of Brooklyn" above water? The trends in residential demographics are already clear. We expect a clean and green post-Sandy Gowanus Canal will be primarily accessed by, and enhance the quality of life of, well-educated neighborhood in-migrants, and longer-term residents of nearby brownstone neighborhoods. These processes are likely to have a disproportionate impact on the Latino population of the community. The increasing density of Gowanus gentrifiers on the banks of the canal will require structural mitigation infrastructure to keep their condos above floodwaters. Such mitigation will be provided by private developers and included in the price of canal-front condos, in a neoliberal model of urban sustainability and climate change adaptation. The wealthy will utilize their private capital to float above the storms, in contrast to the working class that was left to pump them out of their basements after Sandy. Some of the broader infrastructure costs will also be borne by governments, which will be pressed to provide services to areas with high real estate values.

Hurricane Sandy could have been a wake-up call for Gowanus and for New York City as a whole. However, despite years of talk and largely superficial nods to urban sustainability and resilience under the leadership of Mayor Michael Bloomberg, the dominance of real estate interests in New York City's growth machine militated against meaningful reconsideration or redirection of its development paradigm. The new administration of Bill de Blasio, who was elected in 2013, may require a higher percentage of "affordable" housing in new developments, but is unlikely to reverse larger trends in housing markets. The premium placed on waterfront property simply overwhelmed concern for the increased frequency of coastal flooding in the Anthropocene, a proposed name for our current period in geological history, where changes by humans are so pervasive. Likewise, major new waterfront developments at Hudson Yards on Manhattan's West

Side, the Domino Sugar site in Brooklyn, and Coney Island proceeded with only minor nods to increased flooding risk. While the logic of adaptation to ecological conditions argues for a staged retreat from coastal flood zones, the logic of capital argues for increased investment in real estate with water views, even if that water is primarily a sewage outfall.

In Gowanus, Sandy-related flooding was arguably less of a hit to the gentrification process than was Superfund designation. After Sandy, the Lightstone Group moved ahead with its condominium complex on the banks of the canal, despite objections from Gowanus's city councilman. Arguing that its initial plans took full account of federal flood prevention standards, the development corporation stated that "the project was designed to exceed federal 100-year storm standards by significantly elevating the development above the 100-year flood plain."[26] That appears to be the primary response of developers bent on capitalizing on waterfront real estate by increasing coastal population density: build but elevate, so as to be able to have enterococcus infested waters wash over and around luxury residential development. Whole Foods opened on a brownfield on the banks of Gowanus in December 2013 with an overt sustainability theme. Shopping for organic kale chips on an urban brownfield in a coastal flood zone next to a combined sewer overflow while parking your car in a solar-lit lot is certainly a "light" version of environmental consciousness.

Similar developments also continue. The $15 billion, twenty-six-acre Hudson Yards project, located along the West Side of Manhattan and largely in Evacuation Zone 1, carries on, albeit with some structural mitigation. The developers' website affirms it is storm-proofed: "With reinforced fuel tank rooms, watertight points of entry, and an elevated switchgear location, the tower would have been open for business through super storm Sandy."[27]

It is the green growth machine, the convergence of state and private capital interests in transforming both the social and ecological landscapes that makes Gowanus greening possible. The support of community groups makes such greening politically viable. The existence of first-wave gentrifying artists makes such greening saleable. And a pool of potential upper-middle-class gentrifiers makes greening the Gowanus profitable. Housing prices in Gowanus rose 52 percent from 2004 to 2012, when Sandy hit, and have shown no sign of stopping since then. Thus, the green growth machine will create economic growth in the short term, with the long-run consequences of social dislocations and displacement of ecologically damaging production to other, less well-regulated locales, as Brooklyn and the United States continue to deindustrialize.[28] The paradox from an environmental justice standpoint is that the next time the Gowanus floods its banks in a major storm event, it will flood a much wealthier and better-educated community, because the economic logic of the urban redevelopment treadmill is stronger than the storm.

Community activism and government policy can reprioritize the course of development to focus on equity. For example, in the Greenpoint neighborhood of Brooklyn, residents have fought for greening to be "green enough" to benefit locals while not drawing gentrifying in-migrants.[29] Likewise, Mayor de Blasio's "One NYC: The Plan for a Strong and Just City" adds equity as a component to earlier sustainability and resiliency plans. These types of actions and policies will not stop the green growth machine, but they might diminish its inequitable impacts.

LESSONS

This study of the persistence of development in the Gowanus Canal, despite its history of contamination and its vulnerability to flooding, yields several lessons that are relevant to other contaminated coastal sites.

- Green growth machines increase the number of residents living in areas vulnerable to climate change events.

Incentives for developing land in places that green gentrification has made newly attractive counteract concerns about hazards. Normal land use planning processes are inadequate for addressing existing hazards or emerging threats related to climate change.

- Economic logic trumps ecological logic in New York City resiliency practices.

Even under city administrations that have prepared elaborate environmental plans, policies in New York assume that development will intensify. This calls into question whether cities can remake their development policies in ways that avoid placing more people at risk.

- Real estate developers and green gentrifiers are attempting to outsmart ecology with structural mitigation, and the next storm will determine the wisdom of that.

Cities have historically been places that reduce some of the risks to humans while creating other risks. Armoring, elevating, and gentrifying the Brooklyn waterfront is another example of a city creating a new kind of risk, in this case, a risk that we have the knowledge to avoid.

NOTES

1. See Sharon Zukin, "Gentrification: Culture and Capital in the Urban Core," *Annual Review of Sociology* 13 (1987): 129–147; and Neil Smith, *The New Urban Frontier: Gentrification and the Revanchist City* (New York: Routledge, 1996).
2. Dorceta Taylor, *The Environment and the People in American Cities, 1600s–1900s: Disorder, Inequality, and Social Change* (Durham, NC: Duke University Press, 2009).
3. Tom Angotti, *New York for Sale: Community Planning Confronts Global Real Estate* (Cambridge, MA: MIT Press, 2008), 108.
4. Julie Sze, *Noxious New York: The Racial Politics of Urban Health and Environmental Justice* (Cambridge, MA: MIT Press, 2006).
5. See Robert J. Brulle and David N. Pellow, "Environmental Justice, Human Health and Environmental Inequalities," *Annual Review of Public Health* 27 (2006): 103–124.
6. Doug S. Massey and Nancy A. Denton, *American Apartheid: Segregation and the Making of the Underclass* (Cambridge, MA: Harvard University Press, 1993).
7. See Robert J. Bullard, J. Eugene Grigsby III, and Charles Lee, eds., *Residential Apartheid: The American Legacy* (Los Angeles: CASS Publications, 1994); and Kenneth A. Gould, "Promoting Sustainability," in *Public Sociologies Reader,* ed. Judith Blau and Keri Iyall Smith (Lanham, MD: Rowman and Littlefield, 2006).
8. Andrew Szasz, *EcoPopulism: Toxic Waste and the Movement for Environmental Justice* (Minneapolis: University of Minnesota Press, 1994).
9. Kenneth A. Gould and Tammy L. Lewis, "The Environmental Injustice of Green Gentrification: The Case of Brooklyn's Prospect Park," in *The World in Brooklyn: Gentrification, Immigration, and Ethnic Politics in a Global City,* ed. Judith DeSena and Timothy Shortell (Lanham, MD: Lexington Books, 2012).
10. Julian Agyeman, *Sustainable Communities and the Challenge of Environmental Justice* (New York: NYU Press, 2005).
11. See Tammy L. Lewis, "Global Civil Society and the Distribution of Environmental Goods: Funding for Environmental NGOs in Ecuador," in *Environmental Justice beyond Borders: Local Perspectives on Global Inequities,* ed. Julian Agyeman and JoAnn Carmin (Cambridge, MA: MIT Press, 2011); and David N. Pellow and Lisa Sun-Hee Park, *The Slums of Aspen: Immigrants vs. the Environment in America's Eden* (New York: NYU Press, 2011).

12. Sarah Dooling, "Ecological Gentrification: A Research Agenda Exploring Justice in the City," *International Journal of Urban and Regional Research* 33, no. 3 (2009) defines "ecological gentrification" as "the implementation of an environmental planning agenda related to public green spaces that leads to the displacement or exclusion of the most economically vulnerable human population, homeless people, while espousing an environmental ethic" (621). Melissa Checker, "Environmental Gentrification and the Unsustainable 'Downside' of Grassroots Environmental Justice Activism" (paper presented at Nature, Ecology and Society Colloquium, CUNY Graduate Center, New York, 2010), uses the term *environmental gentrification* to mean "ecological sustainability with the rehabilitation of low income neighborhoods."

13. Ioan Voicu and Vicki Been, "The Effect of Community Gardens on Neighboring Property Values," *Real Estate Economics* 36, no. 2 (2008): 241–283.

14. Jonathan D. Essoka, "The Gentrifying Effects of Brownfields Redevelopment," *Western Journal of Black Studies* 34, no. 3 (2010): 309.

15. Harvey Molotch, "The City as Growth Machine," *American Journal of Sociology* 82, no. 2 (1976): 309–332; and John R. Logan and Harvey L. Molotch, *Urban Fortunes: The Political Economy of Place* (Berkeley: University of California Press, 1987).

16. Allan Schnaiberg and Kenneth A. Gould, *Environment and Society: The Enduring Conflict* (Caldwell, NJ: Blackburn Press, 2000); Gould, "Promoting Sustainability"; Kenneth A. Gould, Allan Schnaiberg, and David N. Pellow, *The Treadmill of Production: Injustice and Insustainability in a Global Economy* (Boulder, CO: Paradigm Publishers, 2008).

17. Ellen Snyder-Grenier, *Brooklyn! An Illustrated History* (Philadelphia: Temple University Press, 1996).

18. Martin V. Melosi, *The Sanitary City: Urban Infrastructure in America from Colonial Times to the Present* (Baltimore, MD: Johns Hopkins University Press, 1999).

19. J. S. Nyman, H. Schwartz, and R. Scanlon, *Reconsidering Gowanus: Opportunities for the Sustainable Transformation of an Industrial Neighborhood* (New York: Steven L. Newman Real Estate Institute of Baruch College–CUNY, 2010).

20. Ikea ultimately located on the waterfront in nearby Red Hook, in a structure built on stilts to vent brownfield gas emissions.

21. Nyman, Schwartz, and Scanlon, *Reconsidering Gowanus.*

22. Steve Dollar, "Superfund Site Morphs into Cultural Hotspot," *Wall Street Journal,* December 15, 2010.

23. Nicole Youngman, "Understanding Disaster Vulnerability: Floods and Hurricanes," in *Twenty Lessons in Environmental Sociology*, ed. Kenneth A. Gould and Tammy L. Lewis, 2nd ed. (New York: Oxford University Press, 2015), 231–245.

24. Corcoran Group Real Estate, accessed January 12, 2011, http://www.corcoran.com.

25. "PlaNYC: A Greener Greater New York," 2007, NYC Mayor's Office of Sustainability, http://www.nyc.gov/html/planyc/html/about/about.shtml.

26. Janet Babben, "Development near Gowanus Canal Moves Forward after Sandy," WNYC News, November 12, 2012, http://www.wnyc.org/story/252523-builder_moves_forward_after_sandy.

27. 10 Hudson Yards, accessed May 29, 2015, http://www.hudsonyardsnewyork.com/office/10-hudson-yards/availabilities/building/sustainability-and-resiliency.

28. Schnaiberg and Gould, *Environment and Society*; Gould, "Promoting Sustainability"; Gould, Schnaiberg, and Pellow, *The Treadmill of Production*.

29. Winifred Curran and Trina Hamilton, "Just Green Enough: Contesting Environmental Gentrification in Greenpoint, Brooklyn," *Local Environment* 17, no. 9 (2012): 1,027–1,042.

8 • BOARDWALKS REBORN

Disaster and Renewal on the Jersey Shore

MARK ALAN HEWITT

The shore. To Native Americans it offered plentiful sources of fish and shellfish. European explorers discovered safe harbors at its extremities and dangerous sandbars along its length. The Dutch avoided it, settling the valleys of two great rivers to the north and south and plying their skills as traders. Swedes and Germans found only pestilence on the lower Delaware before William Penn claimed land upstream of the bay for his "green towne." For nearly two centuries, the New Jersey colony was a barrel tapped at both ends, by New York and Philadelphia, with a breadbasket in the middle. Rivers, not open seas, were its economic lifeblood. It was not until later that the Jersey Shore became recognized for what it would become in the popular imagination: the prototypical resort for millions of bourgeois Americans—beginning with small-town resorts such as Cape May City and Long Branch, and eventually spreading out to encompass nearly every acre of shoreline and many acres of former wetlands.[1] The Jersey Shore hotel and the many boardwalks became anchors and archetypes of Jersey Shore tourism, followed by vast numbers of small bungalows built by those wishing to spend more time at the shore at lower cost.[2]

Less acknowledged by promoters of tourism is that by developing this dynamic region, people have invited disasters from fires and storms.

When disasters have struck, cultural and economic forces have encouraged rebuilding. Without reflection on the long history of failed attempts to control beach topography and engineered attempts to protect against fire and storm, these cultural and economic forces appear likely to encourage rebuilding in vulnerable ways after Hurricane Sandy as well.

The ephemeral nature of shore real estate was evident from the very first attempts to colonize New Jersey's beautiful coastline. Formed by glaciers millions of years ago, Jersey's sandy plateau continually shifted between ocean and land, forming barrier islands that mitigated water action and moved sediment up and down the coast, a natural process known as *littoral drift* that has shifted the position, length, and depth of barrier islands and the shoreline itself for millennia. Before extensive building occurred in the mid-nineteenth century, storm-driven currents, floods, and both wave and wind erosion were governed by natural patterns that maintained beach dunes and sand levels in a relatively stable ecosystem.

With settlement, humans have chosen to erect defensive breakwaters, sandbars, jetties, and other kinds of mitigating structures. Orrin Pilkey and Katharine Dixon have demonstrated that all large-scale human intervention to control beach and barrier island ecosystems has been a failure when compared with areas in which no building occurred.[3] The development of beach resorts behind these engineered structures invited patterns of destruction that have plagued the storm-prone Atlantic and Gulf coasts for decades. Each round of destruction has prompted more intervention, including very expensive beach replenishment. Why have these areas persisted in rebuilding following such devastation? The obvious answer is money, derived primarily from property and tourism. As geographer Charles Stansfield has written, "The linear resort complex we know as the Jersey shore is a product of supply, demand, and technology."[4] Tourist demand is fueled through changing cultural trends. The shore has seemed always to resurrect itself, if not always in the same spot and in the same fashion. Erase one boardwalk and two would be built. Once patterns of human habitation have been established, they are hard to roll back, even in the face of climate change and environmental crises.

This chapter links the cultural history and trends of shore development to prevailing socioeconomic responses to Sandy, highlighting in particular the shore as a symbol of a uniquely American form of wealth and power over the past century and a half. It offers a review of how the pattern of rebuilding came into being and why the history of architecture and the built environment should not be overlooked while we search for solutions to climate change and environmental stresses in coastal areas. This story begins with the earliest developments near Philadelphia, because that is where the magic of the Jersey Shore first struck Europeans.

CAPE MAY AND THE SOUTH

As Philadelphia grew to become the first large commercial city in the American colonies, its increasingly sophisticated citizens began to look for respite from the crowded metropolis. Cholera epidemics were regular reminders of the unsanitary conditions prevalent in the city. Sea air was a recommended tonic. Though "Cape Island" began as a fishery, as early as 1801 a Philadelphia journalist could call the sandy peninsula "the most delightful spot the citizens can retire to in the hot season."[5] Following the War of 1812, industrious Philadelphians began building boarding houses at this southernmost point in New Jersey, in a small village named for a Dutch Captain Mey. Its fortunate location allowed both shore and bay breezes to ventilate the island during humid stretches in late summer.[6]

Location was a key factor in development there. Its beaches were a short distance by water from three states. By 1823, there were six boarding houses in the village and one proper hotel, all dingy and brown.[7] By the 1850s, Cape May offered what was to be a hallmark of Jersey Shore tourism: excess, size, and overwhelming hype, centered at that time on the gargantuan, white, and clean Mount Vernon Hotel, 306 feet in length along the seaside, with two 506-foot-long wings extending back from the main facade.[8]

A familiar hazard was fire. The first devastating blaze to claim Cape May's central district began in a shop on the last night of August 1869.[9] Probably arson, the fire destroyed all of Washington Street between Ocean and Jackson Streets, extending to the water's edge, including the magnificent United States Hotel. In a pattern that was to be repeated several times, the town's leaders were quick to dismiss the devastation as a minor impediment to growth and immediately set about rebuilding the town center more opulently than before. The city installed a water system, pumps, and a fire brigade to prevent future disaster. But a small fire that began less than a decade later, on November 9, 1878, eventually burned the entire thirty-five-acre central district to the ground.[10] Though the fire-fighting system existed, supplied by seawater and well pumps, "at the close of each season the pumping equipment was disconnected to prevent damage from freezing. Had this not been the case, the fire probably would have been extinguished at its origin."[11] With one catastrophic event, Cape May lost its preeminent position as a seaside resort, although it continued to attract visitors by trumpeting its quiet, balmy environment throughout the nineteenth century.

After developers built a rail extension on the north side of the island, connecting it to the peninsula, more resort havens emerged on the county's barrier islands to cater to the early-twentieth-century visitors from Delaware, Maryland, and Pennsylvania. Thus Wildwood, Avalon, and Stone Harbor grew to become more successful than Cape May, the old queen of resorts. Motels, the preferred accommodation of the doo-wop era, sprang up by the dozens, while Atlantic City's large hotels attracted more and more middle-class families. Cape May began a period of decline that lasted until the late 1970s.[12]

None of the peninsula towns were prepared for the Great Atlantic Hurricane that struck along a large portion of the East Coast, including the Jersey Shore, on September 14, 1944. Despite enduring a storm called "the worst that has ever struck the southern tip of New Jersey," and dealing with wartime shortages, Cape May "began at once the task of reconstruction, rebuilding as permanently and as extensively as possible."[13] According to city-issued publicity material, the town was back to normal in only a year, with stronger breakwaters and storm-resistant construction.[14] The same indomitable attitude was in evidence twenty years later, though the damage and social disturbance was far greater, when the Ash Wednesday Storm hit the center of the New Jersey coast on March 8, 1962.[15] President John F. Kennedy visited the affected areas shortly afterward, pledging federal assistance for shore infrastructure improvements. Newspapers estimated the total damage in Cape May County at more than $200 million, with one thousand homes lost.[16]

By that time the U.S. Army Corps of Engineers had been managing shore erosion and building storm barriers along most of the New Jersey Shore for decades. Scientists had also studied the climate and ecology of the beaches, salt marshes, and estuaries up and down the coast. In 1962, the *Philadelphia Evening Bulletin* reported that the Army Corps "presented their advice to shore dwellers on how to deal with their sometimes enemy, the Atlantic Ocean: It is this: Retreat and take up positions behind new fortifications. It is very simple advice."[17] Foreshadowing current-day resistance to transformational change in the wake of Sandy and climate change predictions, leaders from the south to the north of the Jersey Shore resolved to resist all assaults on their way of life and livelihood.

LONG BRANCH AND THE NORTH

Long Branch vied with Cape May as the leading shore resort for more than a century and has laid claim to being the earliest combined hotel and bathing establishment in the state.[18] Settlers made their homes there in the seventeenth century, and an early Quaker meetinghouse was constructed as their place of worship. New York City's wealthy society later made it one of the most fashionable summer places in the new nation.

The author Henry James and painter Winslow Homer were the most articulate artists in conveying the beaches and dunes of Long Branch. The two women in the foreground of Homer's 1869 painting (in the Museum of Fine Arts, Boston) are leaning forward into a strong breeze, parasols ready, looking at the bathing pavilions and beach below a high bluff. The evidence of human invasion is patent, as we see a building precariously attached to the side of the dune, with an artist engaged in the act of painting on its balcony, at left.[19]

James came later, in 1903, when the dunes were adorned with a "chain of villas" so striking he compared them to a pearl necklace. Trains and ferries from New York had established the northern Jersey Shore as a fashionable, indeed necessary, playground of the rich. In *The American Scene*, James introduces the reader to the expensive symbols of achievement, lacking any long-term purpose or ambition, writ with a crude force on the landscape of New Jersey: "[E]xpensive as we are, we have nothing to do with continuity, responsibility, transmission, and don't in the least care what becomes of us after we have served our present purpose," he wrote. "The highest luxury of all, the supremely expensive thing, is constituted privacy."[20] Americans were privatizing a natural resource that, perhaps more than any other on earth, would resist such attempts.

Luxury and exclusivity were the draws of Long Branch and its satellite towns. The city of Long Branch was often compared to the gambling scene in Monte Carlo during the post–Civil War years. Elberon had its own exclusive hotel designed by William Potter and Charles McKim of McKim,

Mead & White, well-known New York architects. Many of the earliest Shingle Style "cottages" were there, catering to the likes of Long Branch's developer Lewis B. Brown, H. Victor Newcomb, and Moses Taylor, designed by McKim, Mead & White between 1879 and 1886.[21] Lamb & Rich, Potter & Robertson, and other prominent New York firms designed buildings in the town, including the 1879 Episcopal chapel, now called the Church of the Presidents for the seven chief executives who worshipped there.

The northern shore towns became known less for their boardwalks than for their gated enclaves and beach clubs. Deal, a smaller and more exclusive enclave, became the showplace for real extravagance among New Yorkers, as wooden cottages gave way to stone castles and chateaux along the lines of Newport, Rhode Island, or Long Island's Gold Coast. While Elberon's fortunes waned in the mid-twentieth century, Deal maintained its cachet, particularly among wealthy Jews, who were often effectively excluded from other spots. Even the conservative religious communities of Ocean Grove and Spring Lake, founded by Methodists and Irish Catholics, respectively, have maintained their insularity and caste-protective boundaries. By contrast, Asbury Park, which developed large hotels and boardwalk amusements after Long Branch dropped in popularity, became the most popular of the day-trip resort towns by the middle of the century. During the latter part of the twentieth century, its association with the Stone Pony nightclub and Bruce Springsteen made it a symbol of rugged individualism and resilience, despite economic troubles that have not ebbed.

In the north, some areas were preserved as parks and nature preserves to benefit the larger public. Sandy Hook, part of the Gateway National Park system, has long enjoyed popularity as a surf fishing and bird-watching haven. Manasquan, Sea Girt, and Belmar have also maintained their scenic beaches, while running popular deep-water fishing tours for serious anglers. Near the wide mouth of the Navesink River, Sea Bright and Rumson developed as bedroom communities for New Yorkers in the 1940s and 1950s. All of these estuary areas have been prone to flooding, tidal shifts, and storm damage.

The effects of Atlantic storms have been less evident in these tidal areas, though more rivers empty into the ocean here than in the south. Both the Great Atlantic Hurricane of 1944 and the Ash Wednesday Storm in 1962 were felt acutely, but the damage was less and psychological effects marginal along this section of the shore. Perhaps because of this, northern communities have been less willing to take new precautionary measures in the wake of the biggest storms on record in August 2011, with Hurricane Irene, and in October 2012, with Hurricane Sandy.

THE BARRIER ISLANDS

Sandy came ashore near Brigantine, but it hit hardest at the most vulnerable section of the New Jersey coastline, where a series of barrier islands protect the mainland from the Atlantic. The northernmost barrier islands protect Barnegat Bay and Little Egg Harbor, with the Barnegat Peninsula (more like an island) on the north, including Island Beach State Park, and Long Beach Island on the south, which includes part of the Forsythe National Wildlife Refuge. Just to the south of these islands are Brigantine Island, also including part of the Forsythe National Wildlife Refuge, followed by the more developed coastal Absecon Island, beginning with Atlantic City and ending with Longport at its southern extremity. The Cape May County barrier islands previously discussed extend south of Absecon Island. All of these islands are part of fragile ecosystems that are home to birds, wildlife, and sea life that help to maintain the coast as a whole.

These barrier islands could not have been developed had it not been for transportation technology that transformed the United States in the early nineteenth century. One of the nation's first railroads, the Camden & Amboy, was built in 1833 to connect northern Jersey's industries with those of the south, and New York with Philadelphia. Two decades later in 1852, Dr. Jonathan Pitney engaged a Philadelphia engineer named Richard B. Osborne to plan a direct rail line from Philadelphia to the shore and obtained a charter for the "Camden & Atlantic Company." Thus began one of the most successful real estate ventures in American history, when on July 1, 1854, the first train arrived on Absecon Island from Camden, New Jersey, and Atlantic City was launched as an enduring tourist resort for Philadelphia's urban middle class.[22]

Virtually everything associated with Atlantic City benefited from hyperbole. As the first nationally prominent resort in the nation, the city drew observers from everywhere, even Europe, as early as the 1870s. Running north and south on the island were two broad boulevards, Atlantic Avenue and Pacific Avenue. Perpendicular streets described the most populous states in the nation, inviting tourists to imagine themselves at home while they vacationed.

Pitney and his partners could not have picked a less propitious spot for their experiment. Visitors found streets piled high with beach sand, and the area, originally a mosquito swamp, was inhabited by more blacksnakes and mosquitos than people. In 1870, the first boardwalk made dune scaling obsolete, but the boardwalk had to be dragged off the beach in the fall and reerected for the next season. After a winter storm in 1884 destroyed the boardwalk, a more permanent structure was built that was five feet above the sand and twenty feet wide.[23] The first hotels in Atlantic City were made of stick lumber, easily destroyed by fires and strong winds. Some were rebuilt, larger and more commodious, every fifteen years or so.[24] In order to gain more advantageous views of the sea, entrepreneurs built long piers extending from the boardwalk, but these soon became so large and closely spaced that the hotel owners revolted.

Only an act of the New Jersey legislature, the Beach Park Act of 1894, brought helter-skelter development to a halt. No more piers were allowed, and hotel owners even agreed to create a small city park on the only open site adjoining the boardwalk, called Park Place.[25] In 1896, the boardwalk was expanded to forty feet in width, and it stretched some four miles along the island. Electric lights became neon lights, and the Boardwalk presaged the Great White Way.[26]

"When one is tired or wants to study humanity, there is no place equal to the Boardwalk," said *Heston's Handbook* in 1897.[27] In its final 1896 incarnation, the Boardwalk has lasted until this day. Just as the current landowners do, the merchants, hoteliers, and speculators in Atlantic City at the turn of the twentieth century continually expanded and rebuilt their establishments in order to meet the demands of the time and to reflect the increasing value of the properties they owned. When the Monopoly board game was first marketed in 1933, with its board spaces named after Atlantic City streets, Atlantic City had long since reached its zenith. It was somewhat revived by the advent of casino gambling in the late 1970s, but competition from other gambling venues in the region has once again undercut the city's tourism economy.

Land value on the Jersey Shore is a highly fluid thing, especially on the barrier islands. The greatest concentration of high-value real estate at risk is in Atlantic City, with billions of dollars of casino and hotel properties.

But in other ways, Atlantic City is merely a miniature version of the shore as a whole, with all of its characteristic problems and opportunities. Like most developments, those on Absecon Island were dependent on transportation corridors: first a railroad and later a speedy highway, the Atlantic City Expressway, which is increasingly inadequate to the task of bringing thousands to and from the casinos.

North of Absecon Island lies one of the most beautiful and least spoiled areas of the New Jersey coast, with the Forsythe National Wildlife Refuge and Island Beach State Park. Like much of the shore, even the narrowest of the barrier islands were first privately owned. Whereas John D. Rockefeller developed a huge estate and a hotel in nearby Lakewood, and George Gould built his own Georgian Court nearby, their colleague Henry Phipps of Bethlehem Steel acquired the entire slender Barnegat Island. It was not until 1953, when the state purchased his land and created a state park in 1959, that Island Beach State Park became a public resource.[28]

During most of the nineteenth century, Ocean County (incorporated in 1850), including portions of the mainland and a barrier island, remained sparsely settled and remote, with Barnegat Bay rarely visited except by fishermen. The shoals and sandbars off the barrier islands were dangerous for mariners, and shipwrecks were common. Among the earliest and most successful lifesaving operations were those perfected to rescue seamen and bathers off of Island Beach itself, instigating the formation of the United States Life-Saving Service in 1848, which later became the U.S. Coast Guard.[29] It was a barren and dangerous place to live.

Between 1900 and World War II, Ocean County's population remained constant, at about twenty thousand. After the war, with the advent of the automobile and the construction of the Garden State Parkway in the 1950s, people began moving to the seaside in great numbers. By 1960, the county had more than one hundred thousand residents. Just as a single rail line had driven the growth of Atlantic City, a single highway made Long Beach Island and Barnegat Island accessible to millions of New Jersey residents. Point Pleasant Beach, Seaside Heights, Ortley Beach, and Beach Haven then became havens for sun worshippers, when these quintessential Jersey vacation spots could be reached easily by car. Increasingly affluent middle-class residents were able to afford second homes, and this was the place to build them during the 1960s through the first decade of the twenty-first century. But as Orrin Pilkey has warned, "You can have buildings or you can have beaches; you cannot have both."[30] Presaging a futile cycle of post-destruction rebuilding that would follow the intense period of growth and change in Ocean County, Ian McHarg's classic book *Design with Nature* condemned development on or even near these fragile ecosystems. In 1983, Rutgers University researchers expressed similar caution in their environmental survey of the entire coast.[31]

Their warnings went largely unheeded. When the Ash Wednesday Storm struck in 1962, southern Ocean County was still relatively undeveloped by present standards. Roads on the barrier islands were flooded and layered with sand. Nearby Ocean City's pier was destroyed by surge damage, and fires broke out in town. Long Beach Island was cut into five pieces, and in New Jersey more than forty-five thousand homes were destroyed or badly damaged.[32] But by 2012, the entire inland edge of the barrier islands had been developed on the scale of a vast Levittown, protected in many places by artificial dunes and beaches replenished by the Army Corps. Suburban beach homes were laid out on grids that resembled the rest of New Jersey, indeed the entire Northeast Corridor. The beaches had become front lawns for tourist developments stretching for miles north and south of the now unique Island Beach State Park. When Hurricane Sandy ravaged these areas, there was nothing left to blunt its fury—little wonder that homes, piers, Ferris wheels, and amusement parks were obliterated from Seaside Heights to Manasquan.

LESSONS

- For regions shaped by nostalgia and sustained by repeated waves of investment, rebuilding is rarely questioned. Disasters have often been treated as events that have been experienced and overcome before, or as rare events, interpretations that downplay coastal risk. Either interpretation encourages promoters to rebuild.

Even following the storm of 1889, Cape May's leaders could boast of their resiliency: "When Cape May is in any trouble or has bad luck we are always ready to face the music, and we don't wish to cover anything up.

In this storm we have come out with the greatest success."[33] But if Sandy does not alter development patterns in New Jersey's key shore areas, what kind of transformational event will? As this brief survey of socioeconomic responses to a century and a half of coastal hazards has shown, residents of the three major regions along the Jersey shore have expressed their attitudes toward the sea in various ways, but always pressed to maintain their coastal enterprises in the face of disasters. In the Delaware basin, sturdy Quakers and Pennsylvania Dutch were undaunted by changing weather and constant threats of fire. In Atlantic City, generations of hotel and spa entrepreneurs built and rebuilt the city as a resort, from the first risk takers who settled the city to today's casino moguls. Their unlikely success encouraged cocky attitudes toward any threat short of the fate of the mythical Atlantis, for which one casino was named.

- Plans for rebuilding focus on human uses and on past stories of recovery, even though a changing climate calls these ideas into question.

When Governor Chris Christie proclaimed the strength of New Jersey's people following the 2012 storm, he was tapping a deep vein of sentiment about rebuilding, fueled by narratives about shore recoveries in popular culture. If indeed narratives of rebirth continue to guide the residents of this hundred-odd-mile string of sandy paradise we call the Jersey Shore, they surely emerged first in the people who came to settle and prosper there. While building their dreams and adapting to harsh conditions in non-temperate months, the first settlers established routines that helped them to persist after disasters visited their home places, time and again. Once these narratives were planted in specific locations, among generations of people looking for a stable, even profitable, way of life, they became more and more real, a part of the everyday, and no longer consciously evident.

Understanding how narratives of rebirth have been integrated into attitudes toward the sea allows for a reframing of these narratives in a new context of ecological resiliency. As Joanna Burger and Larry Niles explain in chapter 5, the resiliency of human communities depends in part on the resiliency of coastal ecosystems. The benefit of hindsight that this chapter provides gives an opening to changing our unconsciously repeated narratives that have led to harmful patterns of human intervention in terms of the built environment. It also provides an opening as we look to a future in which we know that whether or not coastal armor or environmental restoration projects are built, there will be future Sandys and devastating fires on this changeable coastline.

NOTES

1. Emil R. Salvini, *The Summer City by the Sea: Cape May, New Jersey, an Illustrated History* (New Brunswick, NJ: Rutgers University Press, 2004).
2. Karl F. Nordstrom, Paul S. Gares, Norbert P. Psulty, William J. Neal, Orrin H. Pilkey Jr., and Orrin H. Pilkey Sr., *Living with the New Jersey Shore* (Durham, NC: Duke University Press, 1986).
3. Orrin Pilkey and Katharine L. Dixon, *The Corps and the Shore* (Washington, DC: Island Press, 1996), 75–103, on beach replenishment.
4. Charles A. Stansfield Jr., *Vacationing on the Jersey Shore* (Mechanicsburg, PA: Stackpole Books, 2004), 14–15.
5. Savani, *The Summer City by the Sea*, 8.
6. George E. Thomas, "Cape May, an American Resort," in *Cape May: Queen of the Seaside Resorts*, by George E. Thomas and Carl E. Doebley, 2nd ed. (Cape May, NJ: Mid-Atlantic Center for the Arts, 1998), 21.
7. Savani, *The Summer City by the Sea.*
8. Thomas, "Cape May, an American Resort," 19–20.
9. Ben Miller, *The First Resort: Fun, Sun, Fire, and War in Cape May, America's Original Seaside Town* (Cape May, NJ: Exit Zero Publisher, 2010).
10. Savani, *The Summer City by the Sea.*
11. Claire E. Lang, "Cape May Ablaze," typescript, 2 pages. Collection of the Cape May County Historical Society.
12. Savani, *The Summer City by the Sea.*
13. National Oceanic and Atmospheric Administration, "The Hurricanes of the 1940s in Virginia and North Carolina," http://www.erh.noaa.gov/akq/Hur40s.htm; *Cape May: A Picture Story of This Resort's Recovery from the Hurricane*, brochure, Collection of the Cape May County Historical Society.
14. *Cape May: A Picture Story of This Resort's Recovery from the Hurricane.*
15. "Devastating Storm Ravages County Resorts," *Wildwood Leader*, March 8, 1962.
16. *Cape May: A Picture Story of This Resort's Recovery from the Hurricane.*
17. "Build New 'Defense Line,' Engineers Advise Shore," *Philadelphia Evening Bulletin*, March 7, 1962.
18. Stansfield, *Vacationing on the Jersey Shore*, 13.
19. *Long Branch, New Jersey*, oil on canvas, 16 × 21.75 inches, 1869, Museum of Fine Arts, Boston, accession number 41.631.
20. Henry James, *The American Scene* (1907; reprint, New York: Scribner's Sons, 1946), 11.

21. See Leland Roth, *McKim, Mead & White, Architects* (New York: Harper and Row, 1983), 68. The initial work was commissioned by Brown and his partner, Louis Sherry, the New York restaurateur.

22. Charles E. Funnell, *By the Beautiful Sea: The Rise and High Times of That Great American Resort, Atlantic City* (New York: Knopf, 1975), 3–11.

23. William H. Sokolic and Robert E. Ruffolo Jr., *Atlantic City Revisited* (Mount Pleasant, SC: Arcadia Publishing, 2006).

24. Ibid.

25. Ibid.

26. Funnell, *By the Beautiful Sea*, 120–121.

27. *Heston's Handbook: Atlantic City Illustrated* ([Atlantic City, NJ: A. M. Heston], 1897).

28. Ocean County Historical Museum, "The History of Island Beach State Park," http://www.islandbeachnj.org/History/History.html.

29. United States Coast Guard, "U.S. Lifesaving Service," accessed June 5, 2014, http://www.uscg.mil/tcyorktown/Ops/NMLBS/Surf/surf1.asp.

30. Pilkey and Dixon, *The Corps and the Shore*, 53.

31. Ian McHarg, *Design with Nature* (New York: Wiley, 1969); Rutgers University Center for Environmental and Coastal Studies, *New Jersey's Barrier Islands: An Ever-Changing Public Resource* (New Brunswick, NJ: Rutgers University Press, 1983).

32. National Oceanic and Atmospheric Administration Eastern Regional Headquarters, "Mid Atlantic Winters," http://www.erh.noaa.gov/lwx/winter/DC-Winters.htm; Nicholas Huba and Kirk Moore, "'62 Storm Effects Still Felt in NJ," *Atlantic City Daily Journal*, March 4, 2012.

33. "The Storm," *Cape May Wave*, September 14, 1889, collection of the Cape May Historical Society.

9 • A SURE/SHORE THING?

Tourism Recovery in New York and New Jersey after Hurricane Sandy

BRIAVEL HOLCOMB

Tourism, a major industry in both New York and New Jersey, was expected to be significantly diminished in the immediate aftermath of Hurricane Sandy, but government and industry efforts to restart business enabled tourists to return to most shore areas by the first two summer seasons after the storm. This chapter describes the nature of shore tourism in New York's Long Island and in New Jersey, discusses policies and market factors affecting the uses of these shore areas, and looks at indicators of whether these areas are rejuvenating or declining after Sandy. By juxtaposing the early post-Sandy marketing campaigns launched by the shore tourism industry and government so as to engage resident loyalties in support of rebuilding at the shore with other, less emotional, attempts to discourage coastal redevelopment, this chapter expands discussion of the socioeconomic reasons why retreat from the shore is not perceived by many as a viable option. Although, theoretically, Sandy could have set a new and different course for tourism development in the coastal regions, in reality the main theme seems to have been to return as far as possible to the status quo ante, to restore the shore to its former glory.

TOURISM ALONG THE NEW YORK AND NEW JERSEY SHORES

Tourism is a changeable business, prone to interruptions by bad weather or bad press and subject to the whims of fashion. One of the few theories in the field of tourism is R. W. Butler's resort life-cycle model, which postulates that tourist resorts start small, grow steadily, experience consolidation, then stagnate.[1] What happens next, rejuvenation or decline, depends on many factors, both local and regional. One could argue that the New York and New Jersey shore, epitomized by Coney Island and especially Atlantic City, has experienced several rounds of decline and rebirth (see chapter 8 by Mark Hewitt).[2]

The shores of both New Jersey and Long Island have comparatively long traditions of recreation, especially for middle-income families for whom a summer sojourn on the sand compensates for living in one of the most crowded parts of the country. The draw of these shore resorts has fluctuated over the years. As jet travel came within financial reach of metropolitan residents in the mid-twentieth century, many abandoned the local beaches for destinations in Florida and the Caribbean. The fashion for lying on the beach soaking up sun was also dampened by the skin cancer scares later in the century. But unlike the Catskill Mountains resorts in New York, which saw tourism permanently decline in the face of jet travel, the popularity of the shore rebounded and has prevailed to date.

The coastal counties of New Jersey and Long Island have experienced especially rapid development since the 1980s, with considerably more land undergoing development for seasonally oriented residential and commercial purposes. The year-round population of at least half the towns in the three New Jersey Shore counties decreased between 2000 and 2010 as year-round homeowners sought to take advantage of this development boom and cashed in by selling their homes to second-home buyers, whose main residence is elsewhere. By 2010, all the towns on Long Beach Island in Ocean County had at least 60 percent seasonal occupancy, and in Harvey Cedars, more than 80 percent of the homes were seasonal.[3] Much of the new construction at the shore is high priced, and households with children are declining in numbers. Rebuilt homes require elevation if in a flood zone, and those that do not conform to revised building codes face large increases in insurance costs or may be uninsurable. Although the fates of individual shore towns and cities may rise and fall, the combination of year-round residents, second-home owners, and beach house renters, with substantial numbers of visitors from nearby, is the fairly settled pattern of the tourism economy on Long Island and the Jersey Shore, a pattern that Butler says is possible for some sites to sustain through the loyalties of visitors over decades.[4]

In New Jersey, tourism ranks as the third-leading industry in the state's overall economy. Tourism in New Jersey generates about $38 billion annually, supports about 300,000 jobs (or 10 percent of the state's employment), and brings significant tax revenues to the state government.[5] In New York State, visitor spending is around $54 billion, supports 695,000 jobs (or 8 percent of the state's employment),[6] and generates nearly $7 billion in taxes. However, with the impact of Sandy being concentrated along the coast, the storm had greater impact on New Jersey's tourism. About 62 percent of all tourism revenues in New Jersey are from three coastal counties (Ocean, Atlantic, and Cape May). In contrast, Long Island, New York State's major coastal region outside of New York City (which captures 64 percent of total tourist spending of that state) garners only about 9 percent of tourist spending.

THE DAMAGE DONE

Early estimates of the cost of damage from Sandy were similar for New York and New Jersey. In November 2012, the month following the storm, Governor Andrew Cuomo released an estimate of $33 billion for New York State, including $19 billion for New York City, and Governor Chris Christie estimated New Jersey's damage at $29.4 billion.[7] Winds damaged structures, and flooding demolished houses, boardwalks, and businesses. Shore towns experienced significant storm surge flooding with resultant damage to infrastructure, furnishings, boats, and recreational equipment. Several shore communities whose activities are centered on boardwalks lost their main attraction. Boardwalks in Seaside Heights and Asbury Park in New Jersey, and in the Rockaways and Long Beach in New York were largely demolished, although (despite publicity to the contrary) Atlantic City's Boardwalk suffered only minor damage and Coney Island's recently rebuilt concrete boardwalk survived. Hurricane Sandy also devastated the region's beaches. In New Jersey, 94 percent of beaches and dunes were damaged, with 14 percent suffering significant loss of dune vegetation and erosion of one hundred feet of beach or more.[8]

Atlantic City, the biggest tourist draw of the New York and New Jersey shore areas, was relatively unscathed by Sandy. There had been a major beach replenishment project the previous summer, with more than $18 million spent to deposit dredged sand on a five-mile stretch of shoreline. The replenishment protected the boardwalk and adjacent casino district, and flooding in Atlantic City was mainly on the bay rather than the ocean. Atlantic City's casinos were closed for five days owing to difficulties in accessing the barrier island and crisis conditions in the city and region, but the casinos were not damaged. A British newspaper (the *Guardian*) inaccurately reported that "the world's most famous gambling destination was stripped of its boardwalk after 'Frankenstorm' swept through town."[9] In fact, only a small portion of the boardwalk at the northeastern tip of the city was demolished. However, the perception of major damage continued. A late November 2012 national online poll found 41 percent of respondents believed that the entire Atlantic City boardwalk was destroyed.[10] For the shore tourist industry overall, the loss or severe damage to many thousands of vacation homes was a major blow. Second homes are typically rented by the week at high rates and generate significant revenues for their owners and communities. In New Jersey it was estimated that 347,000 homes were damaged, including 22,000 that were made uninhabitable. In New York State, 10,000 homes were destroyed, mostly along the shore.

FOREWARNED IS NOT FOREARMED

The tourist shores of Long Island and New Jersey are typified by engineered beaches and intensive settlements of cottages and houses, a combination that has attracted more people to the storm-prone coastline. Prophetically, a 2010 report, *The Shore at Risk*, had detailed threats to these shore beaches prior to Sandy.[11] The wide, sandy beaches of Long Island and New Jersey have long been artificially nourished with sand, or even fully replaced artificially after storms. Little of the shoreline is natural in the sense of being unaltered by human agency.[12]

Beach replenishment and hardening of the coastline with groins, jetties, bulkheads, and riprap has been practiced at considerable expense but limited long-term benefit.[13] Despite frequent calls from experts on coastal processes and on climate change to reduce investment in shoreline preservation (see, for example, Orrin Pilkey's plea),[14] support for beach nourishment continues, with the federal government covering an average of 65 percent of the cost. From 1995 to 2002, New Jersey was the state with the highest level of beach nourishment appropriations.[15]

RECONSTRUCTION, NOT REVISION

Although conservationists, academics, and nature lovers argued for making the shore more sustainable by reducing human footprints and restoring natural ecosystems, except in isolated cases the official call was to rebuild the shore stronger, higher, and as profitable.[16] The tagline of the tourism industry's early post-Sandy marketing campaign, "New Jersey: A State of Resilience," seemed largely to be interpreted as an effort to return the state to pre-storm conditions rather than to make it less vulnerable to future storms. The lyrics of the "Stronger than the Storm" campaign song imply that New Jerseyans will overcome setbacks and remake the state even stronger than before.

In the wake of the storm, a business nonprofit organization began a campaign with the slogan "New Jersey: A State of Resilience." Advertisements featuring shovel-wielding workmen and maps of New Jersey were placed in Union Station, Washington, DC, during the presidential inauguration, in New Orleans during Mardi Gras and the Super Bowl, and on the sides of buses in various cities. The campaign was funded by businesses and was primarily intended to attract investments to New Jersey. Subsequently, a major marketing campaign was undertaken to support the recovery of the state's tourism industry. That campaign, with the slogan "Stronger than the Storm," was paid for largely from $25 million of disaster recovery funding authorized by the federal government. The campaign, much to the chagrin of Democrats, used Governor Christie, a Republican then running for reelection, and his family in the campaign videos. It was aimed mainly at residents of New Jersey and included frequent messages on television and radio, advertising in print, and various promotional events like kite flying and sand castle contests.

New York Times columnist David Carr summed up the prevailing attitude in the summer of 2013: "People don't want the Jersey Shore to be rebuilt better than ever: they mostly want it be the same as it ever was."[17] Recognizing that the beach itself required maintenance, he wrote about the sound of an offshore rig pumping sand back onto the beach: "[T]hat chug of renewal, of man pushing back on what nature had done, came to sound comforting as well."[18]

The U.S. Army Corps of Engineers planned to replace 27 million cubic yards of sand, but some experts worried that making the shore more defensible will simply encourage continued development in inappropriate locations. Sandy beaches are, however, often considered the most iconic attraction of the Jersey Shore. While the beaches themselves do generate direct income in the form of beach fees, their greater importance is their ambience of relaxation, health, family enjoyment, and similar positive images. With beach fees typically around $70 for a season badge or $5 for a daily one, most municipalities do not make significant revenue in any year, although summer 2013 badge sales were especially slow because of lower numbers of visitors.[19] So, even though the beaches are undoubtedly more expensive to maintain (both in sand replacement and garbage collection) than the revenue they generate, beaches are the backdrop for most other shore expenditures—from rentals to food to entertainment. It is lost revenue from these secondary but larger expenditures that threatens the tourism industry.

While beach replenishment got under way rather quickly, the repair or rebuilding of homes, despite governmental funding assistance, was delayed by slow insurance payments and confusion over new regulations pertaining to official flood maps. In the immediate aftermath of Sandy, the Federal Emergency Management Agency released preliminary new maps of flood hazards, which included significant areas designated as in the V, or velocity, zone. Homes in the V zones were at the highest danger of flooding from wave action associated with storm surge and either had to be raised on pilings above the levels required for flooding alone or their owners would be charged very high insurance premiums under revisions to the National Flood Insurance Program adopted by Congress just prior to Sandy. In June 2013, revised FEMA maps were issued that reduced the V zones, although some residents of the originally designated danger zones had already elevated their homes on pilings, at considerable cost. Local governments reconsidered zoning and building codes in the aftermath of the storm, leading to delays in decision making about reconstruction or abandonment. This was further complicated by the schedule for release of FEMA's permanent flood insurance maps (explained in chapter 11, by Mariana Leckner and colleagues).

THE FIRST POST-STORM SUMMER SEASON

The summer 2013 rental market at the shore showed reasonable strength, as people responded to marketing and acted out of loyalty and affection for the shore. In both New York and New Jersey, repeat summer visitors who return year after year to the same town, and even the same house, are typical seasonal visitors. The 2013 annual report issued by New Jersey's Division of Travel and Tourism found that the tourism industry "proved to be resilient despite the effects of Hurricane Sandy, a cooler than normal spring, and the federal government shutdown."[20] Visitor spending posted a 1.3 percent increase statewide over the previous year. In Ocean County, New Jersey's worst-hit area, visitor spending declined only 2.3 percent. Cape May County's tourism revenue increased by 2.3 percent and Monmouth's by 4.9 percent.[21] Construction and investment in tourism-specific facilities jumped 24 percent as recovery from Sandy progressed.

Polling carried out in February 2013 found that among New Jersey residents, there was little change in their intentions of visiting the shore. Asked whether respondents expected to spend more, less, or about the same amount of time at the shore compared to the previous year, 64 percent said "about the same" while 13 percent said "more" and 20 percent said "less."[22] By June, another poll found that 96 percent of those who typically visit the Jersey Shore planned to do so again that summer (see chapter 4, by Ashley Koning and David Redlawsk for details).[23] In-state visitors to the shore typically have shorter stays and are less lucrative for the tourism industry than those visiting from out-of-state, though some pay property taxes on their vacation homes. A small but profitable shore market, especially for southern New Jersey, is Canada, a long day's drive away. Canadian tourism showed a surge in summer 2013, driven by enhanced marketing efforts and a weak

American dollar. Canadian visitors typically stay longer and spend more than domestic tourists, often taking home household goods purchased at lower tax rates than in Canada.[24] Fortunately for them, the southern shore area was less damaged by Sandy.

While Sandy is probably not responsible for much of the decline in visitation to Atlantic City, it is a convenient scapegoat. An article noted a 30 percent decline in the number of bus passengers arriving in Atlantic City between February 2012 and the same month in 2013, and attributed blame to lingering effects of the storm.[25] By the beginning of December 2012, Atlantic City had lost nine conventions accounting for twenty-three thousand hotel room reservations and $31 million in spending, according to the president of the city's Convention and Visitors Authority.[26] But ironically, hotel occupancy in the fourth quarter of 2012 hit a record high in Atlantic County as a result of the need for temporary housing for relief workers and displaced residents. A study that used data on luxury taxes, hotel taxes, and parking fees in Atlantic City found improved collections after that, in the first quarter of 2013, even as gaming revenue declined.[27] A bright spot for the city was the return, after a six-year hiatus in Nevada, of the Miss America pageant in September 2013, on a three-year contract.[28]

These short-term trends in Atlantic City unfolded against the background of longer-term, non-climate-related problems, the largest being competition from other, new, gambling destinations in adjacent and nearby states. Atlantic City's fate is somewhat distinct from that of the smaller shore towns. In his outline of the life-cycle of resorts, Butler accurately predicted that Atlantic City would likely decline if similar gambling attractions were built nearby.[29] Gaming in Atlantic City was made legal in 1978 but was confined within the state to that destination. A report based on 2008 data found that Atlantic City was responsible for a third of all tourism in New Jersey and was second only to Las Vegas in attracting visitors.[30] However, the opening of casinos and gaming in adjacent Pennsylvania and growth of the industry in Maryland and New York has resulted in a sharp decline in visitors to Atlantic City. In June 2015, Governor Christie declared that he supported the construction of new casinos in the northern part of the state if a referendum passed to allow it.[31] In June 2013, Atlantic City's casinos experienced a 12.6 percent decline in revenue from the previous year. The newest casino, Revel, fresh from bankruptcy, posted a 22.6 percent decrease in 2013.[32] An effort to bolster the casinos' fortunes was the introduction (in late 2013) of online gambling. Gamblers are required to visit Atlantic City to register at a casino but can subsequently bet from the comfort of their own couches. This improved casino revenues, but it may further reduce Atlantic City's non-virtual visitor numbers. In 2014, four casinos closed, including the Revel.[33] As the summer 2015 season began, three casinos were in bankruptcy, and one was being picketed by workers.[34] Atlantic City remains a struggling town. Its median household income in 2010 was just over $30,000 compared with a state median of $71,000, with 23 percent of the city's residents living in poverty.

As of summer 2013, much of the shore of both states had to a large extent returned to near normality, although there were, of course, towns and neighborhoods still showing evidence of the storm's devastation. On Fire Island, New York, where two hundred homes were washed away and the Atlantic Ocean overcame fifteen-foot dunes to inundate other homes, business owners organized a $100,000 public relations campaign to bring back the usual seventy-five thousand visitors to an island with only three hundred permanent residents. A local grocer reopened in January to serve the legions of construction workers who converged on the island for repair work.[35] By May, the mayor of Fire Island's Ocean Beach held a press conference to announce that the island "is back." In February 2014, the *Financial Times* announced that the Rockaways, though "brought to its knees by Hurricane Sandy," is now "reborn" as the "new Brooklyn."[36] Equivalent scenes and events were replicated along the shore of both states. And the spirit of making do while the sun shines was evident. The Jet Star roller coaster at Seaside Heights, which fell into the ocean and remained a symbol of Sandy's impact on the Jersey Shore for six months, was removed, but a nearby amusement park defiantly installed a new ride named Super Storm.

LESSONS

- The fate of individual shore towns may rise or fall, but tourism on the shore in the United States is likely to be sustained.

Overall, the outlook for tourism recovery post-Sandy is relatively rosy, given the tourism industry's short-term focus. Sandy may be seen as a minor perturbation in a few years. As Butler, reviewing worldwide trends noted, "[E]ven where specific events have caused tourism to decline in their wake, such as the attacks on tourists in Egypt, visitor numbers have generally been restored to pre-attack levels within 2–3 years."[37] One could perhaps argue that given the relatively short expected life span of beach homes and the current availability of grants and loans to rebuild, it may be rational to rebuild or renovate for twenty years of future pleasure. Despite predictions of sea level rise that would affect the region's beach vulnerability, official policies in both states encourage residents to remain living in the most damaged counties.

- Market and policy incentives for developing the shore are very high in locales that have tourism potential, overwhelming planners' attempts to discourage coastal development.

While there are various efforts to encourage more resilient building and infrastructure, the delays in issuing new rules, competitions for resilient design, and procrastination in addressing the need for new building regulations tempered somewhat the rush to rebuild after Sandy.[38] Obviously, there are often excellent reasons to delay redevelopment, and it could be argued that governmental and nonprofit hesitation and even procrastination in restoring the status quo can be the right strategy, especially when there are serious questions about the sustainability of a rebuilt shore. But for people whose livelihoods depend on attracting tourists to the shore, delays in returning to normal can be financially devastating. There is, of course, much thought and attention being given by many who seek more resilient planning and architecture in the face of climate change and probable sea level rise, but the shore tourist season is only three months long and invites shorter-term planning than that for, say, Manhattan.

While environmentalists would presumably like to see more promotion of nature tourism (the Cape May peninsula attracts three hundred thousand birders annually, who are above average in age, income, and expenditure), and more of what Ian McHarg, using the Jersey Shore as a case study, called designing with nature,[39] as long as government is willing to heavily subsidize beach replenishment and investors are willing to build on pilings or avoid flood zones, it appears that people from metropolitan areas will continue to emulate their forebears and seek a patch of paradise on earth each summer.

NOTES

1. R. W. Butler, "The Concept of a Tourist Area Cycle of Evolution: Implications for Management of Resources," *Canadian Geographer* 24., no. 1 (1980): 5–12.
2. Charles Stansfield, "The Rejuvenation of Atlantic City: The Resort Cycle Recycles," in *The Tourism Area Life Cycle: Applications and Modifications*, ed. R. W. Butler (Clevedon, UK: Channelview Publications, 2006), 287–305.
3. Christie Rotondo, "Is Jersey Shore Real Estate Simply Unaffordable for Young Families?," Press of Atlantic City.com, April 15, 2015, http://www.pressofatlanticcity.com/news/is-jersey-shore-real-estate-simply-unaffordable-for-young-families/article_9dad654a-e0a7–11e4–9d81–8374e26108da.html.
4. Butler, "Concept of a Tourist Area Cycle of Evolution," 10.
5. Tourism Economics, "The Economic Impact of Tourism in New Jersey 2014," accessed June 1, 2015, http://www.visitnj.org/new-jersey-tourism-research-and-information.
6. Tourism Economics, "The Economic Impact of Tourism in New York 2012 Calendar Year," accessed June 1, 2015, http://www.governor.ny.gov/sites/governor.ny.gov/files/archive/assets/documents/tourism/nys-tourism-impact-2012-v1.0.pdf.
7. Editorial, "Hurricane Sandy's Rising Costs," *New York Times*, November 27, 2012, A32.
8. Jenny Anderson, "Rebuilding the Coastline, but at What Cost," *New York Times*, May 18, 2013.
9. "Sandy Aftermath: Atlantic City Counts the Cost," *Guardian*, October 30, 2012.
10. Heather Haddon, "Shoring up N.J. Tourism," *Wall Street Journal*, December 1, 2012.
11. Tony Dutzik and Doug O'Malley, *The Shore at Risk: The Threats Facing New Jersey's Coastal Treasures, and What It Will Take to Address Them* (New Brunswick, NJ: Environment New Jersey Research and Policy Center, 2010).
12. NOAA Coastal Services, "Beach Renourishment: The Lessons from One Long Island Community," July/August 2002, http://www.csc.noaa.gov/magazine/2002/04/beach.html; John Seabrook "The Beach Builders: Can the Jersey Shore Be Saved?," *New Yorker*, July 22, 2013, 42–51.
13. Karl Nordstrom and Nancy Jackson, "Temporal Scales of Landscape Change Following Storms on a Human-Altered Coast, New Jersey, USA," *Journal of Coastal Conservation* 1 (1995): 51–62.
14. Orrin Pilkey, "We Need to Retreat from the Beach," *New York Times*, November 15, 2012, A36.
15. NOAA, *Beach Nourishment: A Guide for Local Government Officials*, accessed March 21, 2014, www.csc.noaa.gov/archived/beachnourishment/html/human/socio/geodist.htm.
16. Mark Tercek, "4 Smart Ways to Rebuild after Superstorm Sandy," *Christian Science Monitor*, November 5, 2012, http://www.csmonitor.com/Commentary/Opinion/2012/1105/4-smart-ways-to-rebuild-after-superstorm-Sandy/Use-data-to-decide-where-and-how-to-build.

17. David Carr, "A View from the Beach Towel," *New York Times*, August 4, 2013, TR6.
18. Ibid.
19. "New Jersey Shore Sees Drop in Beach Revenue after Sandy," CBS Philly, July 27, 2013, http://philadelphia.cbslocal.com/2013/07/27/new-jersey-shore-sees-drop-in-beach-revenue-after-sandy/.
20. Tourism Economics, "The Economic Impact of Tourism in New Jersey, Calendar Year 2013," accessed March 21, 2014, http://www.visitnj.org/sites/default/master/files/2013-nj-economic-impact.pdf.
21. Ibid. The Sandy Hook unit of the Gateway National Recreation Area, located at the very northern tip of the Jersey Shore and operated by the National Park Service, is a very popular tourist destination that was closed by the federal government shutdown that year.
22. "Rutgers-Eagleton Poll: Sandy Not Spoiling Shore-Goers' Plans," Eagleton Institute press release, February 18, 2013.
23. "Bright Outlook on First Day of Summer: Two-Thirds of New Jerseyans Plan to 'Go down the Shore' This Season," Rutgers-Eagleton Poll press release, June 21, 2013.
24. Michael Miller, "Weak American Dollar Draws Canadians to Vacation in South Jersey," Press of Atlantic City.com, August 4, 2013, http://www.pressofatlanticcity.com/news/press/atlantic/weak-american-dollar-draws-canadians-to-vacation-in-south-jersey/article_0465758e-fca2–11e2–9f25–0019bb2963f4.html.
25. "Effects of Sandy Blamed on Drop in Atlantic City Tourism," NJ.Com, June 17, 2013, http://www.nj.com/atlantic/index.ssf/2013/06/effects_of_sandy_blamed_on_drop_in_atlantic_city_tourism.html.
26. "After Hurricane Sandy: How Hard-Hit Destinations Are Working to Win Group Business," Meetings & Conventions, accessed August 12, 2013, http://www.meetings-conventions.com/News/Features/After-Hurricane-Sandy.
27. Joshua Burd, "Study: Tourism Indicators Find Atlantic City May Have Finally Shaken off Sandy," NJBiz.com, June 11, 2013, http://www.njbiz.com/apps/pbcs.dll/article?AID=/20130611/NJBIZ01/130619938/&template=printart.
28. Suzette Parmley and Jacqueline L. Ugo, "Ready for the Closeup," Philly.com, August 12, 2013, http://articles.philly.com/2013–08–12/news/41292218_1_miss-america-parade-boardwalk-atlantic-city-alliance; Jennifer Brogan, "Millions in State Money behind Miss America's Return to Atlantic City," Press of Atlantic City.com, August 11, 2013, http://www.pressofatlanticcity.com/news/breaking/millions-in-state-money-behind-miss-america-s-return-to/article_b798bccc-140c-5f13-a1be-3a51fdba9786.html.
29. Butler, "A Tourism Area Cycle of Evolution," 9.
30. Michael Lahr et al., *The Contribution of the Casino Hotel Industry to New Jersey's Economy* (New Brunswick, NJ: Center for Urban Policy Research, Rutgers University, 2010).
31. Matt Friedman, "Push to Get North Jersey Casino Gambling Question on the Ballot Intensifies," NJ.Com, May 28, 2015, http://www.nj.com/politics/index.ssf/2015/05/campaign_to_get_north_jersey_casino_gambling_quest.html.

32. Suzette Parmley, "A.C. Casinos' June Revenues down 12.6%," Philly.com, July 12, 2013.
33. Suzette Parmley, "Casino Closings in Atlantic City May Diminish N.J.'s Tourism Economy," NJ.com, September 21, 2014, http://www.nj.com/atlantic/index.ssf/2014/09/casino_closings_in_atlantic_city_may_diminish_njs_tourism_economy.html.
34. Wayne Parry, "Pickets, Bankruptcy Greet Atlantic City Casinos for Unofficial Start to Summer," NBC10.com, accessed June 1, 2015, http://www.nbcphiladelphia.com/news/business/Atalantic-City-Casino-Workers-Summer-Icahn-Stockton-304654381.html.
35. "NY's Fire Island, Slammed by Sandy, Says It's Back," *Wall Street Journal*, May 10, 2013.
36. Emily Nathan, "The Rockaways Reborn," *Financial Times*, February 22–23, 2014, 9.
37. Richard Butler, "Tourism in the Future: Cycles, Waves, or Wheels?," *Futures* 41 (2009): 346–352.
38. See, for example, *Fact Sheet: Rebuilding after Sandy* put out by the State of New Jersey and dated June 2013 or the "Rebuild by Design: Hurricane Sandy Regional Planning and Design Competition," which issued its design brief on June 21, 2013, and announced winning designs in May 2014, with implementation to occur after that point. Or consider the FarRoc competition for a resilient Rockaways, NY, which held its inaugural design workshop in August 2013. See http://www.farroc.com.
39. Ian L. McHarg, *Design with Nature* (Garden City, NY: Natural History Press, 1969); Lena Lencek and Gideon Boster, *The Beach: A History of Paradise on Earth* (New York: Viking, 1998).

10 • LOCAL FISCAL IMPACTS OF HURRICANE SANDY

CLINTON J. ANDREWS

On Ash Wednesday in 1962, a winter storm hit the Eastern Seaboard with unusually high tides and towering waves that took twenty-two lives, destroyed or damaged fifty thousand buildings, and caused $1.3 billion in damage, measured in year 2000 dollars.[1] New Jersey's Monmouth, Ocean, and Atlantic Counties absorbed two-thirds of the damage, and until the arrival of Sandy in late 2012, the 1962 storm was perceived as New Jersey's worst storm. Yet every year the region experiences coastal flooding, and globally such problems occur almost daily. Noteworthy coastal flood years just for New Jersey since 1962 include 1971, 1976, 1979, 1984, 1985, 1987, 1991, 1992, 1999, 2004, 2010, 2011, and 2012.[2] Nationally, after adjusting for inflation and removing Hurricanes Katrina and Sandy as outliers, the direct economic losses caused by flooding have grown exponentially at more than 1.7 percent per year from 1903 to 2012.[3]

Three things have changed in the half-century since the Ash Wednesday storm that have increased our exposure to coastal risks. First, the federal government has gone into the flood insurance and beach replenishment businesses in a big way. Second, in response, building has dramatically intensified in the most vulnerable coastal areas, with real estate values on the Jersey Shore increasing by two orders of magnitude in value from 1962 to 2012 after adjusting for inflation.[4] Third, sea-level rise as a result of global climate change and land subsidence along much of the East Coast has accelerated. At Sandy Hook, New Jersey, relative sea level measured by sediment cores was about 7 meters (22 feet) lower five thousand years ago than it is today, yielding an average rate of sea level rise of 1.4 mm/year (0.5 feet/century).[5] The observed rate of rise for the past half-century has averaged 3.9 mm/year (1.2 feet/century), with further acceleration expected.[6] Similar values are seen elsewhere on the Eastern Seaboard. As Hurricane Sandy has confirmed, a perfect storm of good intentions, unintended consequences, and increasing risk is the result. Sandy has not transformed the way municipalities manage their physical and fiscal risks. Yet the analysis presented here shows that under these three trends, approving the rebuilding of houses after a storm may not benefit municipal tax revenues in the way that officials expect.

This chapter analyzes the fiscal effects of several federal policy scenarios to investigate how real property markets and municipal budgets in coastal communities are adapting and are likely to continue adapting to the threats associated with storm events and accelerating sea level rise. These particular threats are not qualitatively different from the flooding risks that coastal communities have always experienced, but climate change is altering their probable severity. With federal encouragement, local land-use planners are now updating their flood-plain maps and associated regulations and incentives. However, the bigger story is about the allocation of risk in a real estate market that is regulated only loosely within a decentralized intergovernmental system. This market dynamic is the engine that drives the tourism-based economies in the United States and the Caribbean in their drive to rebuild quickly after storms (for further discussion of this dynamic in the United States, see chapter 9 by Briavel Holcomb, and for treatment of the trend in the Caribbean, see chapter 3 by Adelle Thomas). The pricing of risk in the real estate market also weighs on local officials when they consider potentially costly decisions to reduce hazards. By focusing on fiscal questions, this chapter quantifies how dramatically misaligned the incentives to both homeowners and municipalities have become with respect to the broader public interest, owing to federal flood insurance and flood mitigation subsidies. People are rebuilding even bigger in harm's way instead of retreating from flood-prone areas.

COASTAL HAZARD MANAGEMENT

Well before Katrina took out New Orleans in 2005, coastal hazard management had become thoroughly intellectualized and institutionalized. Standard training materials of that era from the Federal Emergency Management Agency (FEMA) present a consequential view of risk that distinguishes between natural hazards such as hurricanes and their consequences for humans by noting that "disasters occur when humans are in the way of the natural hazard."[7] They remind us that we have chosen to get in harm's way: eight of world's ten largest cities are coastal; nineteen of the twenty most densely populated counties in America are coastal, as are eighteen of the twenty wealthiest counties; 53 percent of the U.S. population lives on the 14 percent of land area that is coastal; coastal states together earn 85 percent of U.S. tourism revenues; and additional economically important activities such as ports, oil and gas extraction, and fishing take place in coastal areas. Despite the warnings, settlement near coasts around the world is only increasing. Living on the coast may be dangerous, but it is incredibly attractive.

Three basic strategies for managing natural hazards are available: preventing the event, preventing adverse consequences, and mitigating the consequences.[8] Preventing the event is beyond the reach of individuals and communities. Preventing adverse consequences is the domain of planners, among others.[9] Mitigating the consequences is the domain of first responders.

Mechanistic concepts of risk as the product of a probability and a consequence have given way to more elaborate causal models that emphasize how human vulnerability mediates between exposure to natural hazards and adverse consequences.[10] George Clark and colleagues observe an "interplay of social and physical factors that produce human vulnerability" so that exposure (the risk of experiencing a hazardous event), coping ability (consisting of resistance, the ability to absorb impacts without loss of function), and resilience (the ability to adaptively recover) co-determine vulnerability.[11] Communities vary both in their physical settings and coping abilities, hence, distributional issues come to the fore.[12] A study of local fiscal impacts can provide salient information about whether a hazard event is likely to overwhelm the local coping ability.[13]

The varied coping abilities of coastal communities have led to increasing roles for hazard management at the state and national levels.[14] Ad hoc federal responses to events such as the Galveston hurricane of 1900, the lower Mississippi flood of 1927, and the New England hurricane of 1938 have slowly given way to a more systematic approach.[15] Thus the Flood Control Act of 1936 empowered the U.S. Army Corps of Engineers to build protective structures for nearly all major rivers in the country. The National Flood Insurance Program in 1968 filled a void caused when private insurers abandoned flood-prone markets following the costly Ash Wednesday storm of 1962 and hurricanes Betsy and Camille. The Coastal Zone Management Act of 1972 (16 U.S.C. 1451 et seq.) fostered better management of the coastal zone, a national resource, by state and local governments. FEMA came into existence in 1979 as part of an effort to coordinate a multitude of disparate federal disaster management activities, and it in turn was absorbed by the Department of Homeland Security in 2001. The Coastal Barrier Resources Act of 1982 and its amendment (16 U.S.C.1531 et seq.) prohibited federal expenditures to expand infrastructure in many coastal locations. The Stafford Act of 1988 (PL 100–707) and its amendments solidified the nation's disaster management framework and imposed a planning requirement on states and localities seeking disaster assistance. Regulations implementing the Clean Water Act placed limits on individual behavior. State and local regulatory tools for coastal areas further limited choices about where and how to build. Supreme Court decisions fine-tuned the limits of these regulatory powers, generally weakening them.[16]

Professional land use planning audiences have already gained much valuable knowledge about the state and local roles in regulating coastal land uses, siting infrastructures, and balancing the interests between economic development and public safety.[17] This chapter follows the strand of that literature that asks whether federal largesse has so skewed incentives that we are building more than we ought to in harm's way, as asserted by a number of scholars.[18] It investigates the local fiscal effects of flood events, whose probabilities are rising because of sea level rise, as a window into the specific ways in which these intergovernmental incentives affect local decision makers. Analyzing the effects of disasters on municipal budgets provides insight beyond simple measures of probabilities and consequences, because the ability of a local government to provide services deeply affects the local quality of life.

METHODS AND DATA

This chapter presents case studies of three coastal municipalities in northern Monmouth County, New Jersey, examining what happened on a parcel-by-parcel basis during a series of recent and hypothesized future flooding events, how property owners and municipal officials responded, and what roles state and federal government agencies played. It uses a fiscal impact analysis framework to track the incidence of costs and benefits of each event, under extant intergovernmental resource flows and two alternative scenarios.

Sea Bright, Highlands, and Middletown, New Jersey, are the case study municipalities. All have repeatedly suffered significant property damage from coastal flooding, but they differ in their physical, socioeconomic, and fiscal circumstances. As shown in figure 10.1, they are located in the northeastern corner of Monmouth County, New Jersey, near Sandy Hook, which marks the entrance to New York Harbor. Sea Bright is a resort village that

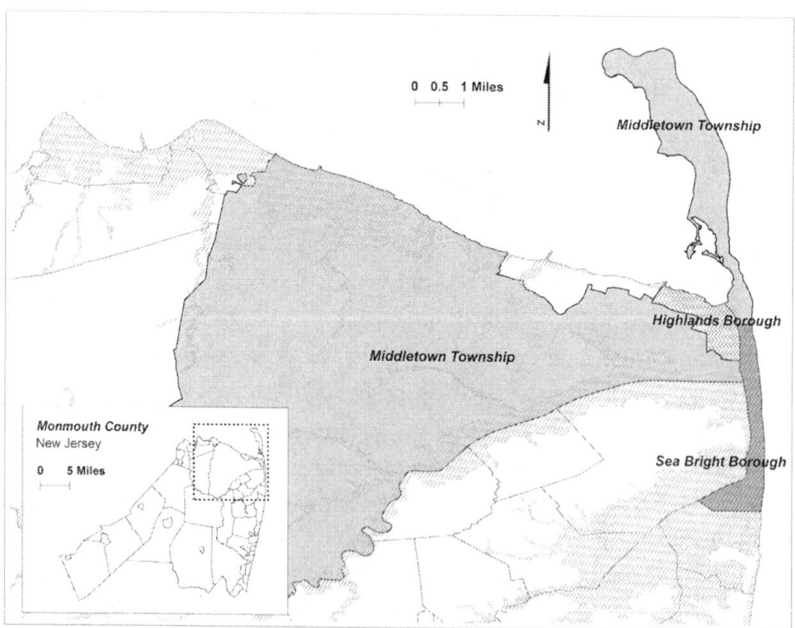

FIGURE 10.1. Map of case study communities (Source: Based on GIS location data from the New Jersey Department of Environmental Protection.)

is highly vulnerable to ocean and bayside flooding, despite being armored by a massive sea wall, with a relatively small year-round population located on a four-mile stretch of a narrow barrier island that is only 500 to 1,100 feet wide. Highlands is a compact, blue-collar borough located on the mainland immediately adjacent to Sea Bright, with its central business district sitting in the floodplain around a centuries-old fishing harbor and its housing stock climbing up a bluff overlooking the ocean. Middletown is a relatively large township in terms of population and land area that sits adjacent to the other two towns, featuring mostly postwar suburban development patterns and large portions that are vulnerable to bay shore and riverine flooding.

Table 10.1 shows comparative 2010 census data for the three municipalities and the overall State of New Jersey, for context. Table 10.2 shows summary information from the National Flood Insurance Program that confirms that even before Sandy, this federal program had paid out far more in local claims than it has received in local premiums since 1978, at ratios of

TABLE 10.1 Descriptions of Case Study Municipalities in Monmouth County, New Jersey

Characteristic/ municipality	New Jersey	Sea Bright	Highlands	Middletown
Geography		*Barrier island, sea wall*	*Bluff and flats*	*Bayshore and riverfront*
Land area (square miles)	7,354.22	0.73	0.77	40.99
Housing units	3,553,562	1,211	3,416	24,959
% built before 1939	18.9%	18.6%	34.3%	11.4%
% owner occupied	66.9%	58.4%	73.4%	86.1%
% for seasonal use	3.8%	24.9%	8.8%	0.8%
Population	8,791,894	1,412	5,005	66,522
Median age, residents	39.0	46.7	44.6	42.5
Median household income	$69,811	$74,236	$75,291	$96,190
Individual poverty rate	9.1%	4.8%	12.3%	3.0%

SOURCE: U.S. Census Bureau, *American Factfinder Profiles of Population, Housing, and Economic Characteristics 2010* (2014).

TABLE 10.2 Flood Insurance Data for Case Study Municipalities in Monmouth County, New Jersey

Characteristic/ Municipality	New Jersey	Sea Bright	Highlands	Middletown
Policies as of 10/02/2012	235,654	1,157	1,128	2,659
Total coverage as of 10/02/2012	$54,200,406,100	$220,350,100	$195,441,900	$706,884,100
Total premiums paid 1978–2012	$211,678,968	$1,154,192	$1,426,523	$2,076,895
Number of claimed losses 1978–2012	111,823	1,296	947	722
Total claims payments 1978–2012	$1,611,317,751	$15,075,885	$10,752,747	$5,778,929
Ratio of payments to premiums	7.6	13.1	7.5	2.8
Number of repetitive loss properties 1978–2012	10,770	130	48	22
Number of repetitive loss claims 1978–2012	NA	347	104	44
Repetitive loss claims payments 1978–2012	NA	$7,282,412	$1,185,092	$633,253
Post-Sandy total claims payments 1978–2014	$5,545,480,808	$79,803,622	$65,694,171	$53,090,788
1978–2014 payments as % of coverage	10%	36%	34%	8%

SOURCES: Federal Emergency Management Agency (FEMA), *Policy & Claim Statistics for Flood Insurance* (2012), accessed June 20, 2014, http://www.fema.gov/policy-claim-statistics-flood-insurance/policy-claim-statistics-flood-insurance/policy-claim-13; Monmouth County, Office of Emergency Management, *Multi-jurisdictional Natural Hazard Mitigation Plan for Monmouth County, New Jersey* (2009), prepared by URS Corp., tables 3a-14 and 3a-15, accessed September 29, 2012, http://www.monmouthsheriff.org/Sections-read-144.html; Paul Weberg, "National Flood Insurance Program (NFIP) and Hazard Mitigation," presentation at Rutgers University, New Brunswick, NJ, February 8, 2012, Federal Emergency Management Administration, Region II, New York, NY.

approximately 3 in Middletown, 8 in Highlands, 13 in Sea Bright, and 8 for New Jersey as a whole.[19]

For each case study, my students and I created a geographic information system (GIS) database with parcel and building footprint layers linked to property tax assessment data, plus natural features, land use categories, detailed terrain mapping using new LIDAR data, and 2011-vintage FEMA floodplain maps. We created a variety of flooding scenarios consistent with probable five hundred–year, one hundred–year, fifty-year, and ten-year storms and identified the affected properties; the one hundred–year simulation is shown in figure 10.2. Using FEMA's Hazus tool,[20] we estimated the dollar value of property damage associated with each flooding scenario. These are underestimates of true damages because we used only the marine flood module of Hazus and did not include riverine flooding, localized flooding from rain, storm surges, or wind damage. Timothy Hall and Adam Sobel estimate the return period for another storm surge following a track like that of Sandy at 714 years (95 percent band 435 to 1,429 years), absent effects of sea level rise,[21] so here we discuss only the Hazus simulations for each of the three case study municipalities for a five hundred–year storm event.

Based on municipal and school budgets for each municipality and associated school districts, we developed per capita estimates of fiscal revenues and expenditures following the methodology of Robert Burchell and colleagues.[22] We multiplied the per capita impacts by the number of households affected by each flooding scenario and the expected number of persons per household. This calculation yielded the net fiscal impacts of each flooding scenario. We investigated each municipality to determine whether there were significant threshold or capacity concerns that would suggest that we should use a marginal-cost rather than an average-cost approach but did not identify any. Finally, we tested the impacts of several alternative state and federal assistance formulas.

We interviewed key stakeholders in each municipality, including local government officials, planners, realtors, insurers, homeowners, renters, and first responders and also spoke with relevant county, state, and federal officials. We also reviewed National Flood Insurance Program (NFIP) claims and payments, other data made available by FEMA, and related literature. These data contextualized our fiscal impact analysis.

Key assumptions in the fiscal impact analysis include the following: During storm events, public safety expenditures, resident relocation costs, and

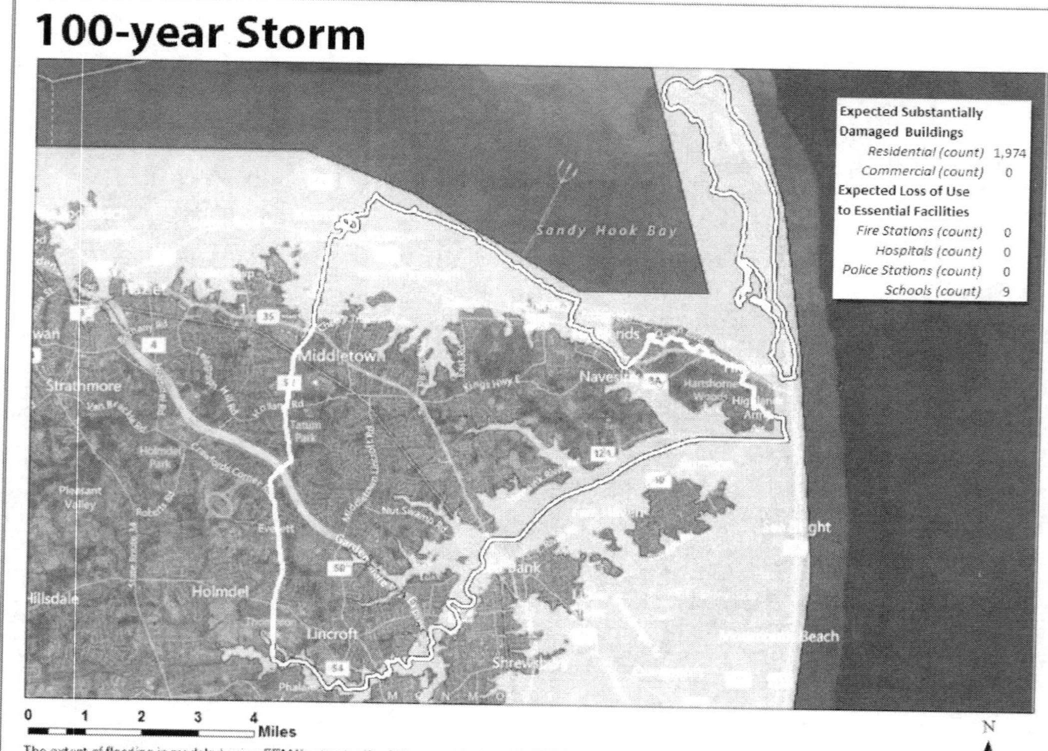

FIGURE 10.2. Flooding scenario for case study communities during a storm with a one hundred–year probability. (Source: Based on Hazus simulation using data from NJDEP and FEMA, shown as a Bing Maps hybrid.)

debris removal costs rise in proportion to storm severity and are paid in the current year. By contrast, we amortize the cost of rebuilding damaged municipal infrastructure over the life of a thirty-year bond. Hazus provides estimates of damage to residential, commercial, and public buildings and infrastructure. Severely damaged residential and commercial buildings leave the property tax base until they are rebuilt. Not estimated are the costs related to loss of productivity under conditions of distress, such the loss of school days or municipal services when municipal facilities are damaged or the loss of business revenue and local purchases when businesses or residences are damaged and their owners are displaced. Also not included are federal grants to localities to replace property taxes, which may be forgiven later.

We examine three policy scenarios, including rebuild in-kind, reduced subsidies, and retreat. Key assumptions are as follows:

Rebuild in kind: We assume that all properties damaged in storms will promptly be rebuilt, some with the assistance of the government, and others through homeowners insurance payments or personal savings. This is the pattern seen in the historic record for the case study municipalities, and it represents the current policy regime. The National Flood Insurance Program continues to cover affected properties (currently it is voluntary and not purchased by all owners of buildings in floodplains), and FEMA continues to cover 75 percent of debris removal costs.

Reduced subsidies: In this scenario we assume that, approximating the now-delayed implementation of the 2013 Biggert-Waters Act, the National Flood Insurance Program actively strives to reduce the imbalance between premiums collected and claims paid in the case study communities. As a result, only 50 percent of property owners receive funds to rebuild. FEMA covers only 25 percent of debris removal costs.

Retreat: We assume that 100 percent of the properties that are substantially damaged (>50 percent of value) in storms do not rebuild. Whether this is the result of FEMA or state buyouts, or simple abandonment, the local result is that these properties disappear from the tax rolls. FEMA continues to cover 75 percent of debris removal costs.

RESULTS

Table 10.3 summarizes the key elements of the municipal budgets for the three case study towns for the years immediately before and after Sandy. Tax rates did not change much, partly because of large intergovernmental resource flows that cover emergency cleanup expenses and temporarily offset local losses in ratables. Whether damaged properties will be rebuilt and reoccupied is an open question at this point. The literature offers some guidance about what to expect. Following Hurricane Andrew in 1992, property values in the hardest-hit areas bounced back quickly in southern Florida's rising market.[23] Following Hurricane Katrina in 2005, property values in areas adjacent to those that were hardest hit rose substantially as the diaspora sought places to live.[24] New Orleans itself suffered low return rates and low real estate prices for several years, and its population remains lower than before the storm, but median home prices in 2014 are higher than they were before the storm.[25] Jacob Vigdor notes that in areas that were losing population before a disaster, housing prices afterward typically have recovered only slowly, if at all, whereas the opposite has been true in areas that enjoy increasing populations.[26] Before Sandy, Highlands was losing population but gaining (smaller-size) households, Sea Bright was losing both population and households, and Middletown had a stable population and number of households.

Table 10.4 summarizes results from a set of fiscal impact analyses of alternative policy responses to a hypothetical storm event approximating Sandy on municipal revenues and expenditures. In every case, the retreat scenario forces the smallest increase in local property tax rate for provision of equal services, because the numerator (expenditures) decreases slightly faster than the denominator (property value). The rebuild scenario, which represents the status quo, produces property tax rates that are higher or equal to those in the retreat scenario, because increased expenditures slightly outpace the increases in property value. The reduced subsidy scenario, in which local jurisdictions bear more of the costs they incur, causes by far the highest local property tax rates, because the proportion of expenditures not offset by intergovernmental transfers increases even as property values do not change.

Historical local property tax rates have been volatile due to changing levels of intergovernmental transfers that are not related to disaster aid or

TABLE 10.3 Municipal Budget Overviews for Case Study Municipalities before and after Sandy

	FY 2012	FY 2013
Highlands		
Municipal property tax	$6,027,505	$6,020,853
Other municipal revenues	$2,067,570	$2,015,192
Emergency aid	$0	$2,081,915
Total municipal revenues	$8,095,075	$10,117,960
Total property tax levy (municipal, school, county)	$15,427,468	$14,923,650
Net value taxable	$606,348,709	$575,346,016
General tax rate per $100 assessed value	$2.55	$2.59
Sea Bright		
Municipal property tax	$3,968,932	$3,471,157
Other municipal revenues	$1,351,052	$833,858
Emergency aid	$0	$3,765,629
Total municipal revenues	$5,319,984	$8,070,644
Total property tax levy (municipal, school, county)	$9,311,615	$8,066,966
Net value taxable	$518,337,818	$447,804,294
General tax rate per $100 assessed value	$1.80	$1.80
Middletown		
Municipal property tax	$43,970,046	$45,051,887
Other municipal revenues	$19,659,597	$22,884,462
Emergency aid	$0	$0
Total municipal revenues	$63,829,643	$67,936,349
Total property tax levy (municipal, school, county)	$208,040,056	$210,053,686
Net value taxable	$9,873,301,487	$9,818,850,218
General tax rate per $100 assessed value	$2.11	$2.14

SOURCE: Municipal budget documents.

NOTE: Middletown is a large and fiscally strong town, and only a small portion suffered damage. Aid was given to individual property owners but not to the town government.

TABLE 10.4 Modeled Local Fiscal Impacts of Responses to Sandy for Case Study Municipalities

Storm event (recurrence period in years)	Municipality	Policy response	% change in expenditures	% change in revenue	% change in property tax rate required for break-even
500	Sea Bright	Rebuild	+9	0	+31
		Reduce subsidy	+30	-11	+46
		Retreat	-46	-21	-29
	Highlands	Rebuild	+58	0	+86
		Reduce subsidy	+49	-15	+104
		Retreat	-6	-29	+31
	Middletown	Rebuild	+16	0	+21
		Reduce subsidy	+14	-7	+28
		Retreat	-5	-15	+10

SOURCE: Andrews et al., "Adapting to Climate Change on Coastal Monmouth County," Spring 2012 Graduate Planning Studio, Rutgers–The State University of New Jersey, http://pppolicy.rutgers.edu/academics/projects/studios/monmoth12.pdf.

insurance payments, state-imposed mandates, and local real estate market dynamics. Sea Bright's equalized municipal tax rate has varied between 0.4 and 1.1 dollars per $100 of assessed value between 1995 and 2012, with annual changes ranging from -23 percent to +11 percent. Annual rate changes have ranged from -25 percent to +15 percent in Highlands, and -23 percent to +11 percent in Middletown during this same time period. General tax rates (shown in table 10.3) including the municipal component plus school, county, and special district taxes, follow the same pattern. The percentage increases in municipal property tax rates shown in table 10.4 for flooding scenarios often exceed the range of historical variation these communities have already experienced, especially for Highlands. The scenarios with reduced federal subsidies and more severe storm events unsurprisingly yield the largest property tax increases.

DISCUSSION

The fiscal impact analysis confirms that increasingly severe storm events are likely to take a severe toll on the coastal communities studied here, no matter which scenario unfolds. Property tax rate increases, an exquisitely sensitive local political variable, are likely to rise above what has been experienced in recent years even if current generous federal and state policies remain in place. A statewide 2 percent cap on increases in effect since 2010 will redirect this imbalance into either reduced services or increased state aid. If higher levels of government pursue policy changes that reduce local subsidies, as seems likely given the parlous state of the budget of the National Flood Insurance Program since its massive payouts for Hurricane Katrina, local municipal positions may become untenable for towns in the most vulnerable locations. Residents in low-lying areas are clearly aware of this possible outcome, given their success in 2014 in persuading Congress to postpone increases in premiums for the National Flood Insurance Program that had earlier been passed under the Biggert-Waters Act, via Public Law No. 113–89 in 2014.

The three case study municipalities differ in important ways. Highlands is the most vulnerable community overall because it has such a small footprint and population and because so much of its infrastructure and year-round housing stock is in the floodplain. Sea Bright is the most physically vulnerable community because it is located on a barrier island that regularly floods, but many of its homes are seasonal, which means that they are good property tax ratables, so the year-round population is somewhat buffered from the fiscal impacts of storm events. Middletown is the least vulnerable community fiscally and physically because it is so large and has less coastal exposure. The bottom two rows of table 10.2 confirm this pattern, with cumulative NFIP payments to date as a percentage of townwide insurance coverage already exceeding 34 percent for Highlands, 36 percent for Sea Bright, and only 8 percent for Middletown. The varied vulnerabilities of these municipalities indicate that part of the route to efficient adaptive responses to climate change involves revisions of local land use plans and zoning ordinances to better internalize the responsibility for preparedness within each jurisdiction.

LESSONS

- Contrary to what local officials might expect, the retreat policy option has superior fiscal performance locally, at least on the coastal plain in eastern New Jersey, and it is also attractive for higher levels of government.

The longstanding FEMA policy of encouraging buyouts of properties making repeated loss claims is an example of this type of policy, although these initiatives have rarely been funded sufficiently to buy out all of the owners willing to sell. Agencies also typically require whole neighborhoods to take the buyout, to avoid leaving holdouts who would still need municipal and utility services. As sea levels rise and storms become more intense, it will become more important to increase the percentage of repetitive-loss properties that do not rebuild. However, no mayor can be seen as giving up on all or parts of his or her town, and it remains almost impossible to talk about the retreat option politically for several reasons, as follows.

- Professional land use planners in municipalities have expertise in balancing hazard mitigation and economic development, but residents committed to living near the shore may not see mitigation regulations and plans as positive encouragements to reduce their own risk.

In an interview before Sandy hit, the mayor of Sea Bright noted that the town flooded approximately six times per year and that residents would often get angry when flooding occurred. The town had changed the zoning ordinance to require houses to be built higher above sea level, which "has been welcomed by most homeowners as they use this opportunity to create a third floor."[27] This is the opposite of retreat.

Some private decision makers did act to reduce their risk, even prior to Sandy. Supermarket owner Scot Bell developed his own nuanced retreat strategy following the 1992 flood by moving his store to higher ground and renting the flood-prone building to other businesses, thereby reducing his risk exposure.[28] However, he did not think that flooding events in Sea Bright had influenced residential property values to a large degree, as "the lure of the ocean keeps them from dropping too much." Prior to Sandy, Tom Thomas, the town's former planner, said: "as far as global warming and all that stuff, I'm going to tell you that nobody pays attention to that."[29]

- Professional planners have not successfully advanced concepts for resilience, and populations are divided about what future they want.

Almost three years after Sandy, although Middletown is fine overall, neither Sea Bright nor Highlands has recovered. Their populations are divided about what future they want, and most improvements are piecemeal, the work of individuals rather than the community as a whole. Properties are elevated, one by one. In response to photo-simulations of alternative rebuilding strategies posted to a Highlands blog, comments were often negative or despairing: "If anyone thinks this will ever happen you can rename the town fantasy land!!!!" "A vision of the downtown as a real destination place . . . that would be fantastic! Too bad I have little faith in the town to be able to accomplish this. I wish we could."[30]

- Without a clear municipal vision and plan for retreat, outside investors looking for wealth-creating opportunities will rebuild, resulting in gentrification.

Fiscal solvency in Highlands and, to a lesser extent, in Sea Bright, depends on either maintaining property tax revenues or reducing public expenditures. Where home sales are taking place, values are rebounding, but many structures are uninhabitable without major reinvestment. Thus, it will often be new people who come in and rebuild to the higher standard that these vulnerable locations require. It appears that the emerging choice is really between gentrification, in places that are attractive to wealthier purchasers, and retreat.

NOTES

Thanks to the students in the studio class who did data analysis that is presented in table 10.4: Marcus Ferreira, Christine Bell, Katherine Nosker, Michael Yaffe, Michael D'Orazio, Kyle Davis, Albert Macaulay, Zhuosi (Joyce) Lu, Judd Schectman, and Brian Gibbons.

1. Gilbert M. Gaul and Anthony R. Wood, "Along the Water: Disasters Waiting for Their Moment," *Philadelphia Inquirer*, March 5, 2000, 1.
2. National Oceanic and Atmospheric Administration (NOAA), "Annual Hurricane Summaries of North Atlantic Storms," accessed June 20, 2014, http://www.aoml.noaa.gov/general/lib/lib1/nhclib/AnnualHurricaneSummaries.html; Rick Schwartz,

"Hurricanes and New Jersey," Hurricanes and the Middle Atlantic States, accessed March 5, 2012, http://www.midatlantichurricanes.com/NewJersey.html.

3. National Weather Service, "Hydrologic Information Center—Flood Loss Data," accessed March 1, 2014, http://www.nws.noaa.gov/hic.

4. Jacqueline L. Urgo and Anthony R. Wood, "Jersey Shore, Forever Redefined," *Philadelphia Inquirer*, March 4, 2012, 1.

5. Norbert P. Psuty and D. D. Ofiara, *Coastal Hazard Management: Past Lessons and Future Directions from New Jersey* (New Brunswick, NJ: Rutgers University Press, 2000).

6. NOAA, "Mean Sea Level Trend: 8531680 Sandy Hook, New Jersey," accessed March 5, 2012, http://tidesandcurrents.noaa.gov/sltrends/sltrends_station.shtml?stnid=8531680; Norbert P. Psuty and Tanya M. Silveira, "Sea Level Rise: Past and Future in New Jersey," technical report available from authors, Rutgers University, New Brunswick, NJ.

7. David J. Brower, Anna K. Schwab, Timothy Beatley, John Bruno, Katherine Eschelbach, and Stephen Meinhold, "Coastal Hazards Management, Instructor Guide," 2006, course materials prepared for the Emergency Management Institute, Federal Emergency Management Agency, accessed April 2, 2012, http://training.fema.gov/EMIWeb/edu/chm.asp.

8. C. Hohenemser, R. E. Kasperson, and R. W. Kates, "Causal Structure," in *Perilous Progress: Managing the Hazards of Technology*, ed. R. W. Kates, C. Hohenemser, and J. X. Kasperson (Boulder, CO.: Westview Press, 1985), 43–66.

9. Timothy Beatley, David J. Brower, and Anna K. Schwab, *An Introduction to Coastal Zone Management* (Washington, DC: Island Press, 2002); David R. Godschalk, Timothy Beatley, Philip Berke, David J. Brower, and Edward J. Kaiser, *Natural Hazard Mitigation: Recasting Disaster Policy and Planning* (Washington, DC: Island Press, 1999).

10. Hohenemser et al., "Causal Structure"; R. Palm, *Natural Hazards: An Integrative Framework for Research and Planning* (Baltimore: Johns Hopkins University Press, 1990); O. Renn, "Concepts of Risk: A Classification," in *Social Theories of Risk*, ed. S. Krimsky and D. Golding (Westport, CT: Praeger, 1992), 53–79.

11. George E. Clark, Susanne C. Moser, Samuel J. Ratick, Kirstin Dow, William B. Meyer, Srinivas Emani, Weigen Jin, Jeanne X. Kasperson, Roger E. Kasperson, and Harry E. Schwarz, "Assessing the Vulnerability of Coastal Communities to Extreme Storms: The Case of Revere, MA, USA," *Mitigation and Adaptation Strategies for Global Change* 3 (1998): 59.

12. S. L. Cutter, "Vulnerability to Environmental Hazards," *Progress in Human Geography* 20 (1996): 529–539.

13. Steven P. French, Dalbyul Lee, and Kristofor Anderson, "Estimating the Social and Economic Consequences of Natural Hazards: A Fiscal Impact Example," *Natural Hazards Review* 11, no. 2 (May 2010): 49–58.

14. W. N. Adger, N. W. Arnell, and Emma L. Tompkins, "Successful Adaptation to Climate Change across Scales," *Global Environmental Change* 15, no. 2 (2005): 77–86; John Posey, "The Determinants of Vulnerability and Adaptive Capacity at the Municipal Level: Evidence from Floodplain Management Programs in the United States," *Global Environmental Change* 19 (2009): 482–493.

15. George D. Haddow and Jane A. Bullock, *Introduction to Emergency Management* (Burlington, MA: Butterworth-Heinemann, 2003); Rutherford H. Platt, *Disasters and Democracy: The Politics of Extreme Natural Events* (Washington, DC: Island Press, 1999).
16. Donna R. Christie and Richard G. Hildreth, *Coastal and Ocean Management Law in a Nutshell,* 2nd ed. (St. Paul, MN: West Publishing, 1999).
17. James C. Schwab, ed., *Hazard Mitigation: Integrating Best Practices into Planning,* Planners Advisory Service Report 560 (Chicago: American Planning Association, 2010).
18. Scholars include Raymond J. Burby, "Hurricane Katrina and the Paradoxes of Government Disaster Policy: Bringing about Wise Governmental Decisions for Hazardous Areas," *Annals of the American Academy of Political and Social Science* 604 no. 1 (2006): 171–191; Robert Deyle, Timothy Chapin, and Earl Baker, "The Proof of the Planning Is in the Platting," *Journal of the American Planning Association* 74, no. 3 (2008): 349–370; Platt, *Disasters and Democracy.*
19. Paul Weberg (Federal Emergency Management Administration, Region II, New York, NY), "National Flood Insurance Program (NFIP) and Hazard Mitigation," presentation at Rutgers University, New Brunswick, NJ, February 8, 2012.
20. Federal Emergency Management Agency, *Hazus: The Federal Emergency Management Agency's (FEMA's) Methodology for Estimating Potential Losses from Disasters,* accessed October 7, 2012, http://www.fema.gov/protecting-our-communities/hazus.
21. Timothy M. Hall and Adam H. Sobel, "On the Impact Angle of Hurricane Sandy's New Jersey Landfall," *Geophysical Research Letters* 40 (2013): 2,312–2,315, doi:10.1002/grl.50395.
22. Robert W. Burchell, David Listokin, and William R. Dolphin, *The New Practitioner's Guide to Fiscal Impact Analysis* (New Brunswick, NJ: CUPR Press, 1985).
23. Federal Housing Finance Authority (FHFA), "Housing Price Index Focus Highlights 3Q2005," accessed March 1, 2014, http://www.fhfa.gov/Default.aspx?Page=193.
24. FHFA, "Housing Price Index Focus Highlights 4Q2005," accessed March 1, 2014, http://www.fhfa.gov/Default.aspx?Page=193.
25. Zillow, Time series from 2004 to 2014 of median sale price of all homes in New Orleans, accessed March 1, 2014, http://www.zillow.com/new-orleans-la/home-values.
26. Jacob Vigdor, "The Economic Aftermath of Hurricane Katrina," *Journal of Economic Perspectives* 22, no. 4 (Fall 2008): 135–154.
27. Interview with Mayor Dena Long, April 19, 2012.
28. Interviewed April 20, 2012.
29. Interviewed April 20, 2012.
30. Community responses to Rutgers Visual Preference Survey, "The View of Highlands Future," *Highlands Blog,* February 16, 2014, http://highlandsblog.me/2014/02/16/the-view-of-highlands-future.

11 • LOCAL RESPONSES TO HURRICANE SANDY

Heterogeneous Experiences and Mismatches with Federal Policy

MARIANA LECKNER, MELANIE McDERMOTT, JAMES K. MITCHELL, AND KAREN M. O'NEILL

Reducing hazards and responding to disasters are activities of risk governance, which is a broader enterprise than government. In the United States, risk governance is a collaborative enterprise that includes several tiers of government, nongovernmental organizations, private businesses, and, especially, victims and non-victims in disaster-affected communities.[1] National and state policies are written to provide uniform standards and a sense of procedural fairness, but these objectives do not easily accommodate the differences within and across coastal municipalities. How did local governments and their residents experience the existing collaborative but fragmented policy system in the aftermath of Sandy? This study of three small municipalities reveals significant disjunctions in the system and examines some implications for changes in policy to address them.

Damage from Hurricane Sandy, and subsequent interactions with federal and state agencies, changed some of the expectations and aims of local government officials and residents in the three towns, but these towns have differed in their abilities to reshape and pursue new aims. Two of the towns were well on their way to rebuilding even six months after the storm, making some adaptations in individual buildings but essentially replicating their earlier footprint. Based on actions so far, it seems unlikely that any of the towns will soon transform their policies to reflect new expectations about risk, even though some officials and residents said they worried more about risk after the storm. The abilities of these particular towns to deal with federal and state agencies have certainly improved. But when the next storm hits, a new set of inexperienced towns, and any new leaders in these three municipalities, will have to begin the learning process.

MANAGING RISKS FOR STORM SURGE IN SMALL MUNICIPALITIES

Many analysts have voiced concerns about the ability of local governments to address climate risks, but these have mostly focused on large cities, especially in poor countries.[2] With the signs of climate change becoming more obvious throughout the world, it is important to identify the abilities of the full range of local governments in coastal areas.

Despite the high visibility of a few big cities like New York and Miami, coastal areas of the United States are dominated by myriad small towns that as a group have limited capacities for governance.[3] These places have widely varying geographic exposures to storms and sea level rise, widely different historic experiences, and widely contrasting response capabilities. Designing and implementing public policies for protecting people, property, infrastructure systems, and ecosystems in these places is difficult because the main policy instruments are managed by federal and state government agencies and large corporations, such as utility companies and insurance firms, whose concerns are often quite different from those of the storm-affected municipalities. Yet it is the local communities that largely determine the success or failure of efforts to create sustainable futures, because they are mainly responsible for implementing relevant public policies.

The difficulties that all local governments experience in managing risk are highlighted in New Jersey towns because of that state's strong home rule tradition favoring local government discretion over land-use decisions. The presumption in New Jersey that municipalities should be highly autonomous makes it especially difficult to coordinate the kind of regional actions that experts say is needed to make coastal communities less vulnerable.[4]

LOCAL EXPERIENCE IN THREE COMMUNITIES: MANASQUAN, OCEANPORT, UNION BEACH

The popular conception of the Jersey Shore is a vacationland comprising small houses, condominium apartments, and hotels set on barrier islands; casinos and shopping venues in Atlantic City; amusement arcades in Seaside Heights and Wildwood Beach; and Victorian guesthouses in Cape May—all facing the open ocean. The reality is much more diverse and also includes former industrial communities on the semi-sheltered shores of Raritan Bay, wealthy residential enclaves on eroding mainland bluffs, suburban tracts on wetland fringed back bays, lagoon developments, commuter dormitory towns along tidal rivers, and quiet rural precincts facing the Delaware Bay.

We chose to study three municipalities in Monmouth County that exhibited varied societal characteristics and a range of experiences with hurricanes. In some ways, Manasquan, Oceanport, and Union Beach are similar to Sandy-impacted towns elsewhere in New Jersey, but in other respects they are distinctive. Insofar as their economies are strongly oriented to amenities, recreational facilities, and real estate transactions, the Monmouth County case study communities have much in common with Ocean County places to the south. But they are on the mainland, not on barrier islands, and so these three towns are generally more elevated. They also contrast with the agricultural shores of Delaware Bay (Cumberland County) and old, established industrial and transportation hubs in Middlesex, Union, Essex, and Hudson Counties to the north.

Seeking information about each town's preparation, experience of the storm, and recovery plans, we conducted two focus groups in each of the three towns, with seven or eight participants in each group. We also interviewed four to five officials, contractors, and staff members from each municipal government who have responsibilities for managing storm surge risks. These included engineers, planners, emergency response managers, council members, and mayors. To capture decisions that were made quite early after the storm, we gathered data between February and April 2013, during a period when one town's flooded borough hall was still being disinfected from mold and another town's borough hall was still housing federal emergency staff.

Manasquan, on the open oceanfront; Union Beach, on the estuarine lands beside Raritan Bay; and Oceanport, on the upper tidal reaches of the Shrewsbury River, all have resident populations of around six thousand and are more or less racially and socially homogeneous, a pattern that is common across New Jersey's smallest shore towns. In Manasquan and Oceanport, family incomes are well above the state average, though neither is a bastion of wealth. Union Beach is less wealthy, but it has about the same percentage of residents living below the state's poverty line as do the other two towns (table 11.1). All three communities contain large areas of land close to sea level and recorded significant damage as a result of storm surge flooding during Sandy. Figures released in March 2013 indicated the number of houses that sustained damage as follows: Union Beach, 1,436 (of 2,269); Oceanport, 409 (of 2,390); Manasquan, 527 (of 3,500). Additional damage to rental properties included Union Beach (279), Oceanport (35), Manasquan (285).[5]

Arriving at high tide, Sandy's storm surge damaged more than eight hundred first and second homes and rental accommodations on or near

TABLE 11.1 Case Study Communities

	Population (2010)	Median household income (2010)	Race/ethnicity	% residents over 65	% residents below poverty line	Exposure to storm surge
Manasquan	5,897	$89,074	96% white	16.3	5.0	High/ oceanfront
Oceanport	5,832	$88,080	93% white	16.1	4.0	Low/riverside
Union Beach	6,245	$65,654	91% white	9.3	5.0	Medium/ bayfront
New Jersey	8,791,894	$71,180	69% white	13.5	9.4	

SOURCES: U.S. Census; New Jersey Department of Community Affairs.

Manasquan's beachfront, knocking some off their foundations and sluicing up to four feet of sand into the bottom floors of others.[6] The surge continued into Raritan Bay, which acted like a funnel, causing water to mount up and pour through the unprotected flank of Union Beach to depths of fourteen feet, damaging more than 1,400 houses, rafting some onto wetlands, and smashing others into second and third rows of homes behind them. Following a more circuitous upstream route from the bay into the Shrewsbury River, the rising waters also swept into Oceanport, which many residents thought of as a river town that was not vulnerable to storm surges, but where more than four hundred houses recorded serious damage.

In a matter of hours, Union Beach lost property that generated almost 10 percent of the tax base that supports its public expenditures. In Manasquan the comparable tax bite was almost 5 percent, and in Oceanport it was under 3 percent.[7] By mid-February 2013, Small Business Administration (SBA) loans of $10 million had been issued to applicants from Manasquan, $13 million in Oceanport, and nearly $20 million in Union Beach.[8] A month later, insurance companies estimated payouts of $21 million to Oceanport, $39 million to Manasquan, and $44 million to Union Beach and adjacent areas of Keyport.[9]

Hurricane Sandy changed all three towns, but not in ways that would lead to a common and transformative vision for reducing future vulnerabilities to climate risks. Each town was damaged differently from the others, owing to its location in relation to the storm track; its combination of landforms, relief, and drainage; the layout of its streets and buildings; and the responses of residents. Each town suffered similar damage to at least one of its neighboring towns, but even this common experience did not inspire much cooperation among town governments.

The knowledge of residents about future storm surge risks was remarkably varied and strongly reflected specific local vulnerabilities. Though often aware of systematic areawide scientific projections of inundation risks compiled by experts in support of the federal government's National Flood Insurance Program, local residents in our focus groups also drew deeply on knowledge borne of personal acquaintance with sites, structures, and signals of risk. In addition, social differences among the three towns affected how they confronted the tasks of recovery and rebuilding. Manasquan and Oceanport, the two higher-income towns, were less financially stressed than Union Beach at the household and government levels and were able to devote more attention to considering their long-term future as waterfront communities.

The complexity, uncertainty, and ambiguity of rebuilding decisions for residents and local officials were dramatically increased by changes in federal policies that were being phased in at the time Sandy hit. A number of new policies had been created after Hurricane Katrina, which hit the Gulf Coast in 2005. The most important of these, and the one that was most talked about by local leaders and residents, was an updated set of Flood Insurance Rate Maps (FIRMs) marking flood and wave height zones for the purpose of setting insurance premium rates. New Jersey had long been scheduled as one of the last states to receive rate map updates, which the Federal Emergency Management Agency (FEMA) had originally scheduled for late 2013 or early 2014, a year or so after Sandy struck.

Instead, FEMA published a set of interim maps a few months after Sandy, together with interim guidelines for rebuilding called Advisory Base Flood Elevations (ABFEs). Towns were told by FEMA that the advisory guidance was not final. Rebuilding would run the risk of substantially higher insurance premiums and other penalties if not consistent with tighter regulations that were likely to be adopted by state and local authorities after final maps and final rebuilding guidance were issued much later. New Jersey quickly incorporated the advisory maps into state regulations, and it has updated state requirements as FEMA has introduced new maps, but this did not fully address residents' uncertainty and was little noted in focus groups.[10] The advisory guidance implied, but FEMA did not require, that owners elevate their damaged single-family houses, if they were in flood or surge zones, by heights often exceeding six feet (there is much less FEMA guidance for multifamily buildings).

Focus group members possessed varying degrees of knowledge about changes in the maps and about the new ABFEs, ranging from vague awareness that such existed, through confusion among the different formulations, to skepticism about their precision and permanence. Residents of Manasquan expressed most familiarity with the relevant language and concepts, while Union Beach participants expressed the least. None of the group members felt the maps and elevation data provided a complete and reliable basis for making decisions about rebuilding. One well-informed participant reported that he was juggling three types of elevation data, his old base flood elevation (BFE), his new ABFE, and an as-yet-undetermined "secret" BFE that might or might not eventually be adopted. The existing maps and elevation requirements were complex enough, but the idea that people would be rebuilding under interim maps that were subject to further revision added immense confusion. None of the maps, old, interim, or permanent, incorporate data about storm surge in New Jersey from Sandy or Hurricane Irene (2011), nor do they incorporate sea level rise projections, a flood-forcing factor that is expected to worsen significantly in succeeding decades.[11] No focus group participant mentioned these deficiencies, a flaw in laypersons' understanding of future storm surge risks that might well have serious repercussions for the long-term sustainability of these communities.

Uncertainty about the meaning and role of new flood risk information was just one of a lengthy list of factors reported by residents in focus groups that complicated the process of recovery and redevelopment. These included at least thirty other factors, ranging from changes of coastal morphology to the reduced attractiveness of damaged communities for visitors. The numbers of those who planned to remain in their homes, to leave, or who were undecided varied across the three communities. Only in inland Oceanport, the least at-risk and least damaged municipality, were there committed leavers. Manasquan, the wealthiest town and the only one of the three towns with a beach section of vacation and rental houses, had the greatest number of committed stayers. Union Beach, the town with the lowest average income and the most damage, had the greatest number who were undecided. These results indicate that residents weigh a wide array of factors beyond criteria such as expected flood heights and exposure to waves that are used to determine risk zones on the federal flood insurance maps.

Local officials and participants in the focus groups also stated that their towns were waiving fees and speeding up building approvals and that some owners were proceeding to rebuild, even before receiving insurance payments. None of our respondents mentioned proposals to pause and reconsider building policies. Activities that FEMA and other agencies would characterize as being part of the later rebuilding stage of disaster response were therefore occurring during the early stages of cleanup and recovery.

LIMITATIONS OF RISK GOVERNANCE AT THE MUNICIPAL LEVEL

Sandy was a differentiating mechanism within and across these communities, as much as a unifying one. Its heterogeneous outcomes are likely to produce sociocultural or political-economic disadvantages that are unevenly distributed. Climate justice advocates point to the harms associated with worsening climate risks that will accumulate for poor people in low-lying tropical countries.[12] Our findings show that new inequalities are now emerging in the wake of Sandy. Senior citizens experienced some of the most severe and lasting impacts. Given the disproportionate numbers of retirees that can be found in many coastal communities, this finding signals the potential for a gradual shift of the elderly away from these areas, likely under financial hardships, especially for those on limited incomes who are unable to refinance long-term new mortgages. Many younger poor and working-class families have also had difficulty returning to their homes after Sandy and will probably have trouble adjusting to losses from climate change.

These trends would alter the social and physical character of some towns and could also affect their viability. Towns such as Union Beach, with large numbers of moderate-income residents and few attractions for the wealthy, may find themselves less and less able to provide basic services. Sharing services with neighboring towns may foster collaboration, but, under current political arrangements, the strengths of one municipality may not make up for the vulnerabilities of its neighbor. We found that officials in municipalities using shared services agreements worried about being burdened by problems of their partners' making after the storm. So inequalities within and across small coastal towns seem likely to increase, posing even greater challenges to national and state policies and private programs that rely on municipalities to reduce overall coastal risks.

Uniform policy approaches, such as federal flood maps, also increased a sense of uncertainty and unfairness because the policies were not consistent with residents' and officials' knowledge about specific local hazards, were based on principles that they did not readily understand, and were still being worked out in the storm's aftermath. In addition, because hazard management can impinge on private property rights, residents' political ideologies as well as their trust in governments affected whether they saw management policies as beneficial or punitive. The formal step-by-step process of assessing damage, determining eligibility for different types of aid, renegotiating existing indebtedness obligations, filing properly documented claims for grants or loans and insurance reimbursements, distributing payments to approved recipients, securing building permits, and engaging contractors was frequently perceived as slow and cumbersome. Uncertainties about the equity of aid were increased by delays in delivering that aid to many residents. Allegations about favoritism in the distribution of aid by Governor Chris Christie's administration occurred well after our study period for these three towns, but statewide perceptions about aid distribution are reported in chapter 4 by Ashley Koning and David Redlawsk.

PROSPECTS FOR THE FUTURE

Varied patterns of responses across the three towns lead us to question whether existing federal and state disaster recovery programs that are organized around discrete stages of disaster are suited to the needs of local municipalities. We found that all three communities made decisions very soon after Sandy to return what Sandy destroyed to its pre-storm state. Even during the cleanup phase, local discussions were shaping the long-term recovery trajectory. Stage models that posit separate or overlapping stages of short-, medium-, and long-term recovery are important as ideal types useful to research analysts, and also for practical reasons, because they affect how federal and state agencies organize their bureaucracies and provide services to municipalities.[13] Federal and state bureaucracies use stage models because the stages reflect legal requirements for the agencies to start and finish tasks, based on formal indicators and on the passage of time. But many residents and officials found these stages bewildering, because they perceived that their town needed particular types of help at particular moments that may not have been recognized by the agencies.

There is already a substantial literature on how governmental institutions can improve reconstruction after disasters.[14] In this study, we learned that laypeople use a plethora of information and ideas about storm surge risks, such as the effects of microscale site features and the location of vulnerable residents. We would add to this literature the recommendation that this local knowledge and experience could help create a more effective risk governance system. Whether and how such a resource might be mobilized and integrated into the existing system, or developed as a separate stream of mediating knowledge, are appropriate subjects for future research. Innovative institutional arrangements, such as the local resilience partnerships advocated by the federal Hurricane Sandy Rebuilding Task Force, also hold out the possibility of strengthening lateral collaboration among municipalities that share similar problems in ways that would complement the vertical collaborative structure (local, state, federal) of the existing disaster recovery system.[15]

We see two likely paths for an individual town that is hit hard by a future storm surge. In a town with the right combination of attractive features, the disaster can become a redevelopment opportunity, where current owners or new investors and developers build houses and condominiums that are larger and safer and that change the character of the town in the process.[16] Other towns that lack such attractions will likely see delays in rebuilding and even abandonment of property, stresses on municipal finances, and other undesirable changes (see Clinton Andrews's chapter 10 for further discussion of these divergent paths). In both types of towns, the elderly, the working class, and the poor will find it difficult to stay.

We also see a special role for federal and state agencies, business firms, and nonprofit organizations to enable more equitable and sustainable outcomes by encouraging towns to adopt a more regional perspective and to engage in longer-term planning. Incentives from federal programs can provide clear signals about the sorts of choices concerning locations and construction of infrastructure and buildings that the national government will encourage or discourage along the coast, but it is equally important to recruit a supportive constituency among coastal residents. An example of an existing federal program that encourages hazard reduction is the Community Rating System, which offsets some NFIP premiums based on local efforts to reduce risks and damages in future storms.

Experts of all kinds favor more active intervention to secure the coast in light of worsening risks. What our study adds to these recommendations is that because some towns are better able to respond than others, public policies on all levels of the federal system will deal better with the inevitable ambiguities of hazard management, disaster response, and disaster recovery if they learn about and accommodate local differences and if they take advantage of the repository of local knowledge that already exists among populations at risk.

LESSONS

Among victims of Sandy and other residents of impacted communities there exists a reservoir of local knowledge about storm surge risks and recovery experiences that has been largely untapped by public decision makers with access to the major hazards and disaster policy and management tools. Better means for identifying these resources and articulating them with disaster recovery actions are needed.

- Local knowledge and experience could help create a more effective risk governance system throughout all levels of government.

Decisions about land development and rebuilding after disasters are concentrated in the hands of local officials, but flood insurance and most of the expertise and funds for assisting rebuilding come from the state and federal levels. Policies at the state and federal levels that are confusing to residents and local officials or that fail to incorporate local understandings of hazards will yield unintended consequences.

- Resilience partnerships may strengthen collaboration between municipalities. Our research of three towns found no new partnerships with neighboring towns.

Leadership from state or federal agencies established well before a storm hits could encourage some forms of collaboration (see chapter 2 by Daniel Hess and Brian Conley about the state government's encouragement for establishing evacuation centers across municipalities).

- If interim guidance is used by federal or state agencies after a disaster, it should be simple, clear, and related to temporary measures, not a short-term replacement for long-term policy.

FEMA had been revising its flood maps for many years and was rolling out new maps state by state. As part of general disaster preparedness, agencies can acknowledge such gaps and make clear to local officials that there would likely be interim measures in the event of a major disaster.

- Regional and long-term collaborative planning at local levels may increase sustainable planning.

Coastal zone management and floodplain management have been enacted in pieces in most states. Incentives for regional approaches and cross-border cooperation, and information about the true costs of risky development, may encourage local governments to incorporate resilience goals into their development policies.

NOTES

This research was supported by grant #1324792, "Post-Disaster Risk Redefinition in Small New Jersey Municipalities during the Initial Recovery Period Following Super Storm Sandy," provided through the National Science Foundation's Infrastructure Management and Extreme Events Program.

1. Federal Emergency Management Agency, *National Disaster Recovery Framework: Strengthening Disaster Recovery for the Nation* (2011), https://www.fema.gov/pdf/recoveryframework/ndrf.pdf.
2. Stephane Hallegatte, Colin Green, Robert J. Nicholls, and Jan Corfee-Morlot, "Future Flood Losses in Major Coastal Cities," *Nature Climate Change* 3 (2013): 802–806; Dirk Heinrichs, Kerstin Krellenberg, and Michail Fragkias, "Urban Responses to Climate Change: Theories and Governance Practice in Cities of the Global South," *International Journal of Urban and Regional Research* 37, no. 6 (2013): 1,865–1,878; P. McCarney, P. H. Blanco, J. Carmin, and M. Colley, "Cities and Climate Change," in *Climate Change and Cities: First Assessment Report of the Urban Climate Change Research Network*, ed. C. Rosenzweig, W. D. Solecki, S. A. Hammer, and S. Mehrotra (Cambridge: Cambridge University Press, 2011), 249–269.
3. For example, of thirty-two tidal municipalities in Monmouth County, New Jersey, Long Branch is the largest, with just over thirty thousand people (year-round population), while twenty-four have populations less than ten thousand, with fifteen of these recording fewer than five thousand. Summer populations can be much higher.
4. Barbara G. Salmore and Stephen A. Salmore, *New Jersey Politics and Government: Suburban Politics Comes of Age*, 2nd ed. (Lincoln: University of Nebraska Press, 1998). It may seem that studying a state with home rule written into its constitution, such as New York, would provide an even more extreme case. In practice, New Jersey's home

rule tradition has created an expectation of local autonomy that matches or exceeds that of constitutional home rule states. Many small coastal towns will face the same issues regarding municipal capacity; see David Wachsmuth, "How Local Governments Hinder Our Response to Natural Disasters," The Atlantic Cities.com, October 28, 2013, http://www.theatlanticcities.com/politics/2013/10/how-local-governments-hinder-our-response-natural-disasters/7362.

5. Colleen O'Dea, "Interactive Map: Assessing Damage from Superstorm Sandy," *NJSpotlight*, March 15, 2013, http://www.njspotlight.com/stories/13/03/14/assessing-damage-from-superstorm-sandy/. Other estimates vary from these loss totals, often adding to them.

6. Damage data compiled by the New Jersey Department of Community Affairs and reported in "Interactive Map: Assessing Damage from Superstorm Sandy," *NJSpotlight*, March 15, 2013, http://www.njspotlight.com/stories/13/03/14/assessing-damage-from-superstorm-sandy.

7. Tax board data for Ocean and Monmouth Counties reported in the *Star-Ledger*, March 31, 2013, http://www.nj.com/politics/index.ssf/2013/03/hurricane_sandy_deals_blow_to.html#incart_river.

8. Sandy Recovery Scorecard, NJ.com, May 23, 2013, http://www.nj.com/hurricane sandy/2013/05/sandy_relief_scorecard.html#incart_special-report.

9. New Jersey Department of Banking and Insurance data reported in "Interactive Map: Sandy-Related Insurance Claims," *NJSpotlight*, April 19, 2013, http://www.njspotlight.com/stories/13/04/17/insurance-claims-from-superstorm-sandy.

10. The final FIRMs were delayed by FEMA's work on Sandy and will likely be released a year or more late, after a period of public comment and community outreach. A subsequent set of interim maps revised the wave height zones but not the flood delineations. The advisory guidelines were adopted as regulations by New Jersey and some municipalities whereas other towns kept their own existing, and sometimes more stringent, requirements (e.g., Oceanport). See, for example, "N.J. Sandy Rebuilding Rules: Go Higher or Pay More," *USA Today*, January 25, 2013, http://www.usatoday.com/story/news/nation/2013/01/25/sandy-rebuilding-flood-maps/1863761; and "In New Jersey, New FEMA Flood Maps Bring Controversy," *JLC: The Journal of Light Construction*, March 25, 2013, http://www.jlconline.com/business/in-new-jersey-new-fema-flood-maps-bring-controversy-and-confusion_o.

11. Matthew J. P. Cooper, Michael D. Beevers, and Michael Oppenheimer, "The Potential Impacts of Sea Level Rise on the Coastal Region of New Jersey, USA," *Climatic Change* 90 (2008): 475–492.

12. Ilan Kelman, "Hearing Local Voices from Small Island Developing States for Climate Change," *Local Environment* 15, no. 7 (2010): 605–619; International Climate Justice Network, "Bali Principles of Climate Justice," August 28 2002, http://www.ejnet.org/ej/bali.pdf.

13. R. W. Kates, C. E. Colten, S. Laska, and S. P. Leatherman, "Reconstruction of New Orleans after Hurricane Katrina: A Research Perspective," *Proceedings of the National Academy of Sciences* 103, no. 40 (2006): 14,653–14,660, www.pnas.org/content/103/40/14653.full.pdf+html; Federal Emergency Management Agency, "Recovery

Continuum: Description of Activities by Phase," *National Disaster Recovery Framework: Strengthening Disaster Recovery for the Nation* (2011): 8, https://www.fema.gov/pdf/recoveryframework/ndrf.pdf.

14. David W. Edgington, "Reconstruction after Natural Disasters: The Opportunities and Constraints Facing Our Cities," *Town Planning Review* 82, no. 6 (2011): v–xi.

15. U.S. Department of Housing and Urban Development, *Hurricane Sandy Rebuilding Strategy: Stronger Communities, a Resilient Region, Hurricane Sandy Rebuilding Task Force,* August 2013, 134, http://portal.hud.gov/hudportal/documents/huddoc?id=HSRebuildingStrategy.pdf.

16. Timothy Beatley, "Planning for Resilient Coastal Communities: Emerging Practice and Future Directions," in *Adapting to Climate Change: Lessons from Natural Hazard Planning,* ed. B. C. Glavovic and G. P. Smith (Dordrecht: Springer, 2014), 123–144.

12 • WATER UTILITIES

Storm Preparedness and Restoration

DANIEL J. VAN ABS

Hurricane Sandy raises many questions about the integrity and viability of water infrastructure in vulnerable coastal areas. This chapter explores the ways in which governments and utilities are responding to the risks and liabilities made clear by Hurricane Sandy, and it assesses whether water utilities will be better prepared for the next storm. The analysis begins with a description of damage to water, wastewater, and stormwater infrastructure in New York and New Jersey, the areas hit hardest by Sandy. I then discuss the relationship of development patterns along the shore to the vulnerability of related water infrastructure. Reports from municipalities and utilities and an informal survey of people in the water utility industry are used to outline the steps that some utilities are taking to reduce risk.

Finally, to consider some of the ways that utilities are responding to these problems, the chapter presents a brief case study of one New Jersey utility that provides drinking water. This utility is improving its routine management tasks with new technologies, an approach that can also help it recover from storms and manage long-term vulnerabilities. The case study focuses on one of the hardest hit areas of the Jersey Shore to assess the initial loss of water utility customers as a result of property damage, and the rates of water demand prior to and following the storm as an indicator of the rebuilding rate.

WHAT HAPPENED?

Hurricane Sandy is well known for its damage to buildings, roads, and electric utilities. Less well known are the damages to water supply, sewage, and stormwater utility systems, collectively known as water utilities. Of the estimated 11 billion gallons of untreated or partially treated sewage released into receiving water bodies of the United States as a result of Sandy, nearly all (93.7 percent) occurred in New Jersey (46.4 percent) and New York (47.3 percent), mostly owing to storm surge damages to sewage treatment facilities.[1] Among the hardest-hit facilities in New Jersey was the Passaic Valley Sewerage Commissioners (PVSC) facility, the fifth-largest in the nation, which released 840 million gallons of untreated sewage to Newark Bay and an additional 3 billion gallons of partially treated sewage to the lower Hudson River. Its treatment facility alone suffered at least $300 million in damages.[2] Another large treatment plant, Middlesex County Utilities Authority, is on high ground and was undamaged, but the main sewage pumping station to the plant was lost to flooding, resulting in 1.1 billion gallons of untreated sewage being discharged to Raritan Bay.

In New York State, the largest discharges were of partially treated sewage from East Rockaway (Nassau County) and Yonkers (Westchester County) facilities.[3] The Bay Park facility in East Rockaway was directly flooded, with damages estimated at $847 million for a facility that has less than one-fourth the capacity of the PVSC plant.[4] New York City's sewage treatment facilities were functional within five days or less, but "ten of 14 wastewater treatment plants operated by the Department of Environmental Protection (DEP) released partially treated or untreated sewage. . . . 42 of 96 pumping stations that keep stormwater, wastewater, or combined sewage moving through the system were temporarily out of service because they were damaged or lost power."[5]

Combined systems using the same pipes for wastewater and stormwater are common in older cities in the Northeast and are prone to being overwhelmed and to allowing untreated water to flow into streams and estuaries during major storms. Coastal sewer utilities that had previously installed submersible pumps suffered less damage (though some had no electricity available and therefore were still not functional). Extensive networks of sewage collection pipelines and storm sewers in many coastal areas, which are difficult to monitor and repair, were also damaged in many locations.

Water supply treatment plants fared better than sewage treatment plants, in large part because Hurricane Sandy's effects were mostly related to storm surge and winds rather than river flooding, which can damage reservoirs and upstream treatment plants, as occurred during Hurricane Irene in 2011. However, water distribution systems in various areas of the New York and New Jersey coast were damaged during Sandy by the same storm surge, sand erosion, and sand overwash that destroyed and damaged so many buildings, sewer lines, and storm sewers.

Damages to water utilities include not only costs for rehabilitation, but also the loss of revenues in the short term and potentially in the long term as well in situations where buildings were destroyed or sufficiently damaged such that demand for water and wastewater services declined. In assessing areas with the greatest loss of customers, New York City found that nearly all of the major building damages were in areas of storm surge and wave action, and were in sections of the city where most of the buildings were constructed prior to 1961.[6] New Jersey exhibited similar impacts.

WATER UTILITIES AND COASTAL DEVELOPMENT

Water supply, wastewater, and stormwater infrastructure are all enablers of coastal development. They do not create development demand as much as they allow for the expression of that demand. Where provision of water infrastructure is technically feasible and legally permissible, construction has occurred and will occur in response to often intense market demands. Where water infrastructure is prohibited or greatly constrained by law or is technically infeasible, little development is possible. For the most part, coastal development patterns in the northeastern United States were established prior to restrictive modern coastal protection approaches, most of which date from the 1970s or later.

Pre-1970s settlement near the shore in the Northeast was therefore driven mainly by culture and economics rather than regulation (as Mark Alan Hewitt discusses in chapter 8). Historic shore towns have been augmented over time by a large expanse of shore bungalows and rental units that appealed to the middle class. Billy Joel's song "Allentown" alludes to this appeal, speaking about the lives of Lehigh Valley residents who worked hard and looked forward to visits to the New Jersey Shore.[7] The result is that coastal development in the mid-Atlantic states is generally dense, with grid street patterns that cover barrier islands and other coastal areas right to the beaches, wetlands, and open waters. Lagoon developments from the 1960s and earlier are prevalent in tidal bays, with the developed lands only a few feet above mean higher high-water (MHHW), a measure of average tidewater levels that can be used to assess the vulnerability of coastal sites.

Disasters do not seem to limit market pressures for long. For instance, development of the Jersey Shore has escalated dramatically since the 1962 Ash Wednesday storm that devastated parts of its barrier islands. Many moderate- or fixed-income families, often retirees or second- and third-generation owners, now own modest homes on very expensive land, much of which is at risk for damage from extreme weather events.[8] This market situation makes it difficult to predict how market demand or regulation will affect future demand for water utility services.

Nearly every developed coastal area in this region is served by three types of public water infrastructure: water supply, sanitary (or combined) sewers, and separate (or combined) storm sewers. Water supply systems may be owned by government entities (for example, municipal departments and utility authorities) or investor-owned companies. Public sewerage systems are nearly all government-owned, and ownership of the sewage collection systems may differ from that of the sewage treatment plant. Storm sewers are nearly always owned by municipal governments or by private developments. Of these, the costs of water supply and sewerage systems are supported primarily by their customers, while the costs of storm sewers, to the extent they are managed at all, are borne by taxpayers or the private owners. How water utilities handle their long-term costs will affect how well municipalities recover from storms, and for utilities owned directly by municipal governments, it will affect the fiscal health of local governments.

SPECIAL ISSUES FACING COASTAL WATER UTILITIES

Coastal water infrastructure is highly vulnerable to major storm damage, especially on barrier islands and within lagoon developments. Storms may erode sandy soils, exposing water infrastructure to wave impacts and structural damage. Storms may shift massive amounts of sand, clogging storm sewers entirely and clogging sewer and water supply lines where these are broken by storm action or building displacement. Sand movement may also bury critical infrastructure such as valves, water and sewage pumps, and manholes and other service access points. Storm surge may inundate the land area, damaging pumps, vastly increasing sewage and stormwater flows, and preventing utility crew access. Water supply and sewer systems are also subject to indirect damages that may be highly disruptive, such as loss of electric power for both treatment plants and pumps in the distribution or collection systems. Finally, for all three types of water infrastructure, the rise of relative sea level will increase routine flooding of developed areas with water infrastructure, posing hazards to both the water infrastructure and the development it serves.[9] As Sandy showed, much of the utility infrastructure in the largest coastal cities is just as vulnerable to storm surge and inundation as is the infrastructure of beach areas. The beach areas, however, have the additional hazard of sand displacement and erosion.

ASSESSMENT OF WATER UTILITY VULNERABILITY

Water utility planning and budgeting for system operations, maintenance, and modification will depend heavily on the knowledge and expectations of system operators and regulators regarding current and future vulnerabilities. Utilities will need to retrofit their physical infrastructure, train their staffs, educate the public, and coordinate with critical service users such as hospitals. In the entire region affected by Hurricane Sandy, New York City has best formalized a major assessment of vulnerability after the storm.

Even before Sandy, New York City had begun to improve flood preparedness and reduce prospective flood damages, in response to its ambitious 2007 PlaNYC initiative, including a revised building code applicable to new buildings. Hurricane Sandy flooded 17 percent of the city, or fifty-one square miles, including all of the city's wastewater treatment plants. Within eight months of the storm, the city released a major report with recommendations for improved resiliency of city facilities and developed areas, with benchmarks of 2020 and 2050 and a price tag of $20 billion.[10] This report represented the final major policy initiative in the administration of Mayor Michael Bloomberg on this issue, with implementation falling mostly to the administration of Mayor Bill de Blasio.

The resiliency report noted that Hurricane Sandy flooded areas far beyond the most vulnerable zones delineated in then-existing Flood Insurance Rate Maps (FIRMs) created by the Federal Emergency Management Agency (FEMA).[11] Sea level rise will increase risks beyond those noted in the FIRMs, with an estimated 24 percent of New York City being within the one hundred–year floodplain by the 2050s.[12] All of New York City's wastewater treatment plants and many of its sewage pumping stations are vulnerable to coastal storms, including risk from sea level rise. Regarding wastewater infrastructure, "the City's potential [financial] exposure was estimated to be $900 million at wastewater treatment plants and $220 million at pumping stations."[13] The report recommends a suite of protective measures, including beach nourishment, dune construction, increased shoreline heights and hardening, offshore breakwaters, wetlands and reefs, levees and flood walls, tide gates, and integrated flood protection systems.[14]

New Jersey lacks an equivalent resiliency plan. Though it released a State Hazard Mitigation Plan in early 2014 that provides general risk assessments and an overview of mitigation efforts, this plan does not set new policy for water utilities.[15] New Jersey does have a robust collection of state-wide data in geographic information system (GIS) database systems, including Land Use/Land Cover, topographic, aerial photo, parcel mapping, and property tax assessment data. The state has also provided municipal planning grants for resiliency planning, which can incorporate these data and interact with local utilities.[16] Assessment of water utility damages and future risks is being performed by individual utilities, assisted by resilience guidance documents related to water utilities prepared by the state government, some portions of which may be incorporated into regulations.[17] These are discussed in more detail later in this chapter.

UTILITY PREPAREDNESS AND PRIORITIES

Water utilities generally face intense pressure to minimize rates, while the aging of infrastructure, regulations, and demands for risk reduction require more investment. In 2013, the New Jersey Climate Adaptation Alliance (NJCAA), a coalition of interest parties, conducted an online survey of professional association members to better understand utility viewpoints.[18] The survey addressed a wide array of issues regarding recent storm impacts (Hurricane Irene of 2011 and Hurricane Sandy of 2012), utility preparedness, asset management, and utility finances.[19] More than one hundred responses were received, primarily from utility managers and staff but also from consultants and contractors, with coastal utilities being more likely to respond than noncoastal utilities.

The most frequently reported impacts from Hurricane Sandy in water utility service areas were long-term power loss, road closures, business closures, flooding, and property damage. The respondents were asked to rate the preparedness of their utilities before Hurricane Irene and after Hurricane Sandy. Those indicating their utilities are now "extremely well prepared" increased from 11 to 26 percent, while those indicating that they were "somewhat prepared" dropped from 28 to 16 percent. A plurality responded that their utilities were "adequately prepared" in both periods (at 42 percent prior to Irene and 44 percent after Sandy).[20]

The greatest concerns of respondents regarding climate change were "increased severity of storms, increased severity of coastal storm surges, physical damage/deterioration to critical utility infrastructure (energy, water, wastewater), increased occurrence/severity of coastal floods, increased frequency of severe storms, and more severe droughts."[21] Despite a fairly strong sense that climate change is occurring and represents a risk, the water utility respondents ranked climate change impacts well below other traditional utility issues such as asset management and regulatory compliance.[22] Yet the types of actions that will meet regulations, yield supportable budgets, and maintain utilities' infrastructure can also help utilities to respond to climate change, as the following brief case study shows.

CASE STUDY: TRACKING THE REBOUND FROM SANDY

An important question for water utilities is how quickly and to what extent pre-storm customer demands will recover to pre-storm levels, which will affect the design of rebuilt infrastructure and ongoing revenues needed to pay for those repairs. Multiple factors affect the rate of redevelopment, including pre-storm development patterns, property owner finances and preferences, government assistance, infrastructure restoration and improvement rates, and how residents perceive recent risk reduction efforts and their own experiences of storms. Damage from Sandy was so severe in some areas that for some water utilities, these trends will remain important business concerns for some time.

Little information has been reported in the literature regarding the trajectory of demand restoration. Future demands from the most damaged portions of the Jersey Shore likely will not match prior demands, for several reasons. First, any redevelopment must incorporate modern, water-conserving fixtures. Second, where new pipelines for water supply have been installed, water losses from pipe leakage and faulty service connections will be lower than before. Third, redevelopment may trend away from rentals and toward wealthier owners of second homes, who can afford to elevate the rebuilt homes. Finally, redevelopment may entail consolidation of some small lots into larger lots, reducing the number of homes. Together, these factors suggest long-term demand will drop in many areas. Conversely, demands in some areas could increase, owing to redevelopment at the higher densities that are necessary to make the development profitable. Also, water demands from larger homes are greater than for smaller homes, and so redevelopment to larger house sizes could increase demand. Tracking actual demand changes through the next five to ten years would help utilities make decisions about their infrastructure and revenues.

A subsidiary of American Water, New Jersey American Water Company (NJ American) is an investor-owned utility that has the largest customer base of any water supply utility in New Jersey. Prior to Sandy, in 2011, NJ American fortuitously had completed a pilot project documenting every water customer connection point in Ortley Beach, a barrier island community within Toms River Township. The project involved using global positioning system (GPS) technology and noting the elevation of each connection valve in a GIS database.[23] The purpose of the pilot project was to determine the costs of expanding this inventory database to the utility's entire service area, to increase the cost-effectiveness of customer service. The data are linked to NJ American's billing information, allowing the company to determine the category, billing status, and water demands of each customer.[24]

These data became highly valuable after Hurricane Sandy. The Mantoloking section of the barrier island north of Ortley Beach (figure 12.1) was dissected by ocean surge and wave action that cut through the island to the

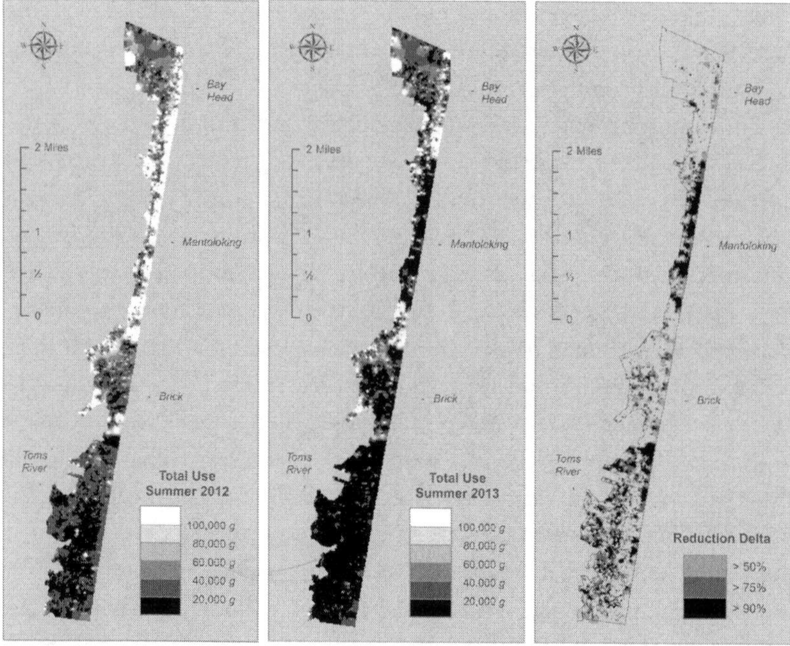

FIGURE 12.1. Water demands before and after Hurricane Sandy, Northern Barnegat Bay Island, New Jersey. (Source: New Jersey American Water Company; used by permission.)

bay, destroying a state highway and all related infrastructure, including water supply lines. Ortley Beach, farther south, was also badly damaged. Many homes were completely destroyed or displaced. Sand overwash buried water service connections several feet below the new surface, often in areas where visual referents were no longer available because the homes no longer existed. NJ American proceeded to extend temporary water lines into the disaster zone, but it also had to cut off water delivery to damaged properties to avoid high-water losses. The GPS coordinates for Ortley Beach allowed field crews to determine precisely where each valve was located, allowing rapid access to the cutoff valves. Where buildings no longer existed or were condemned, NJ American tracked connections using the database to avoid billing inactive customers. By comparison, in the absence of GPS data for the remainder of the damaged areas, response times were much slower.

NJ American used this wealth of data to assess the pre-storm (summer 2012) and post-storm (summer 2013) density of connections and water demands. As a control, similar data were tabulated for Ocean City, a municipality in Cape May County, New Jersey, that experienced minimal damages from Hurricane Sandy. Table 12.1 and figure 12.1 from NJ American show the near-total cessation of water demands in some areas that had relatively high demands prior to the storm. Water demand reductions in excess of 90 percent on figure 12.1 indicate the areas of greatest storm damage, such as Mantoloking. As noted in table 12.1, even at the municipal scale, demand reductions exceeded 40 percent for both residential and commercial customers in the service area, and it exceeded 60 percent for residential customers in the hard-hit areas of Mantoloking and Toms River (primarily Ortley Beach).

By comparison, in figure 12.2 and table 12.2, Ocean City shows relatively little change in water demand between the summer of 2012 (prior to Hurricane Sandy) and the summer of 2013 (after Hurricane Sandy).

Figure 12.3 and table 12.3 show the differences in storm damage between the two areas based on the level of water supply service connections that were logged as inactive in February 2014. Ocean City had almost no inactive water connections. By contrast, at least 13 percent of the residential connections are listed as inactive in each of the Barnegat Bay Island municipalities

TABLE 12.1 Northern Barnegat Bay Island, NJ, Water Demands by Class, Summer 2012 and Summer 2013

Municipality	Billing class	Summer 2012 ('000s gallons)	Summer 2013 ('000s gallons)	% reduction
Bay Head Borough	Residential	56,206	44,172	21
	Commercial	7,969	8,035	-1
Berkeley Township (partial)	Residential	3,297	1,590	52
	Commercial	290	260	10
Brick Township (partial)	Residential	55,472	29,505	47
	Commercial	3,230	1,422	56
Lavallette Borough	Residential	8,162	4,608	44
	Commercial	504	270	46
Mantoloking Borough	Residential	50,755	19,417	62
	Commercial	1,046	111	89
Toms River Township (partial)	Residential	116,004	45,451	61
	Commercial	13,589	3,862	72

SOURCE: New Jersey American Water Company.

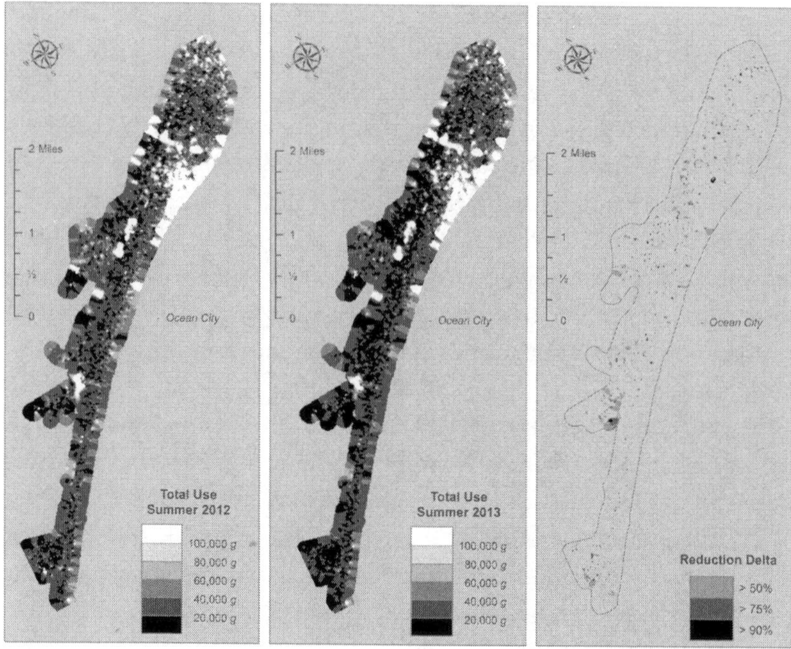

FIGURE 12.2. Water demands before and after Hurricane Sandy, Ocean City, New Jersey. (Source: New Jersey American Water Company; used by permission.)

(or sections thereof), with Mantoloking and the barrier island sections of both Brick and Toms River having more than 40 percent inactive connections. These values may understate the damages, as many homes were flooded but not destroyed, and so retained active connections. Information systems like this can help utilities to project ongoing demand trends and to improve responses after disasters, aiding their responses to climate change.

TABLE 12.2 Ocean City, NJ, Water Demands by Class, Summer 2012 and Summer 2013

Billing Class	Summer 2012 ('000s gallons)	Summer 2013 ('000s gallons)	% reduction
Residential	371	312	16
Commercial	181	150	17

SOURCE: New Jersey American Water Company.

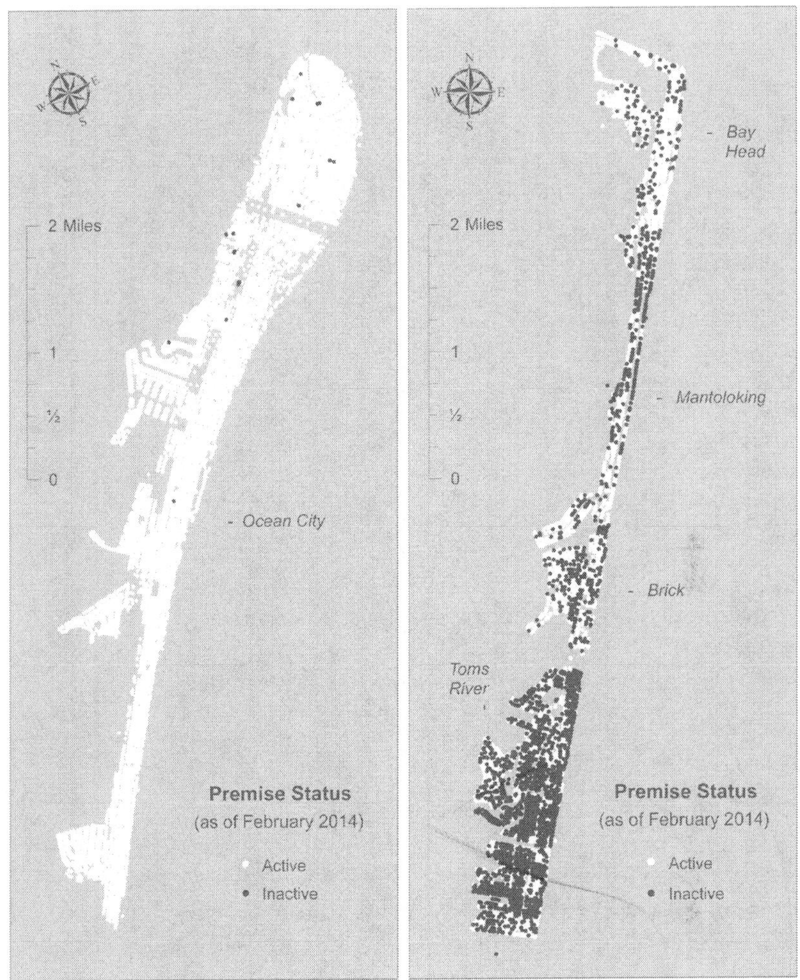

FIGURE 12.3. Comparison of inactive water connections, Ocean City and Northern Barnegat Bay Island, New Jersey. (Source: New Jersey American Water Company; used by permission.)

ARE UTILITIES PREPARING FOR COASTAL STORMS?

Federal agencies such as the U.S. Environmental Protection Agency (EPA) were highlighting potential water utility risks related to extreme weather events prior to Hurricane Sandy.[25] In response to Sandy, the federal Hurricane Sandy Rebuilding Strategy called for all federal infrastructure decisions and grants to incorporate future risks, including sea level rise, into infrastructure

TABLE 12.3 Status of Residential and Commercial Water Supply Connections, Northern Barnegat Bay Island and Ocean City, NJ, February 2014

	Commercial connections			Residential connections		
Municipality	*Status*	*Number*	%	*Status*	*Number*	%
Bay Head Borough	Inactive	7	14	Inactive	124	13
	Active	42	86	Active	852	87
Berkeley Township (partial)	Inactive	0	0	Inactive	34	34
	Active	11	100	Active	65	66
Brick Township (partial)	Inactive	2	11	Inactive	561	41
	Active	16	89	Active	814	59
Lavallette Borough	Inactive	0	0	Inactive	92	29
	Active	2	100	Active	230	71
Mantoloking Borough	Inactive	7	64	Inactive	228	43
	Active	4	36	Active	305	57
Toms River Township (partial)	Inactive	3	5	Inactive	2,862	45
	Active	62	95	Active	3,487	55
Ocean City	Inactive	3	0	Inactive	11	0
	Active	2,489	100	Active	12,212	100

SOURCE: New Jersey American Water Company.
NOTE: Percentages are rounded to the nearest digit.

designs, affecting funding from FEMA, the EPA, and the U.S. Department of Housing and Urban Development (HUD).[26] The President's Climate Action Plan incorporates this policy and mandated other actions to reduce future losses related to sea level rise and other climate change impacts.[27] HUD adopted rules that in most cases prohibited use of HUD funds in coastal high hazard areas.[28] Where federal funds are provided to recipients through state programs, the federal requirements cascade through to the local level, policies that in the long run will reduce utility exposures to risk.

Differences between preparations for future storms by New York City versus those of small New Jersey shore towns are representative of experiences along the urbanized coastline of the United States. New York City operates both water supply and wastewater utilities, and intends to address its system vulnerabilities through a combination of top-down actions, including:[29]

- Adopt a wastewater facility design standard for storm surge and sea level rise through 2050;
- Harden pumping stations and wastewater treatment plants to address inundation issues;
- Explore alternatives for the Rockaway Wastewater Treatment Plant, the most heavily damaged plant;
- Prevent power losses through on-site generation and other measures; and
- Reduce combined sewer overflows with green infrastructure, to reduce stress on the systems, and create separate storm sewers, to divert storm runoff to separate outfalls.

In contrast to New York City, New Jersey has many independent water supply and wastewater utilities, often under separate ownership within individual municipalities, emphasizing the importance of state government actions for reducing storm damages to utilities and reducing vulnerability overall. For example the New Jersey Environmental Infrastructure Finance Program focuses on low-interest loans and principal forgiveness (that is, capital subsidies) for restoring damaged facilities and for natural hazards risk mitigation.[30] And NJDEP released four guidance documents for water supply and wastewater utilities in 2014 regarding emergency response preparedness/planning, infrastructure flood protection, asset management, and other critical planning steps and is considering revising regulations, providing clearer requirements and also linking utilities' compliance to their eligibility for funding. The flood protection guidance incorporates requirements for federal funding such as "elevating critical infrastructure above the 500-year flood (i.e. 0.2% annual flood event) elevation, if feasible, or floodproofing."[31] However, the guidance does not incorporate sea level rise, and current NJDEP regulations have less stringent requirements where no federal funding is involved.[32] Utility managers' qualitative responses to the NJCAA survey discussed above, utility funding requests, and interviews with New Jersey utility managers indicate that the major adaptive responses to extreme weather events focus on improved auxiliary power, elevation of sensitive equipment, replacement with submersible pumps, and floodproofing. The need for better risk assessment information was a frequent concern.

OTHER REGIONS, OTHER STORMS

Water utilities in many coastal areas of the United States and other nations have been damaged by hurricanes and typhoons over the decades. Understandably, most attention is focused on mortality, injuries, and building damages, but water utilities were likewise damaged in places including New Orleans (Hurricane Katrina in 2005), Galveston (Hurricane Ike of 2008 and the Galveston Hurricane of 1900) and the Gulf Coast barrier islands (Hurricane Frederic, 1979, and Hurricane Camille, 1969, among others). In recent years, information on water infrastructure has been more comprehensive, such as that available for Hurricane Katrina, indicating extensive water damages to pumps and electrical systems, inoperable water supply and wastewater treatment facilities, and broken pipes.[33] Likewise, a FEMA report about Hurricane Ike noted that "water and wastewater lines will have to be replaced in areas where the storm reconfigured the state's shoreline."[34] However, information available on these and similar storms focuses on early damage assessments and initial repairs, not on the long-term pace of infrastructure reconstruction.

CONCLUSIONS

Hurricane Sandy in 2012 followed closely on the heels of Hurricane Irene and Tropical Storm Lee in 2011, with their river flooding impacts in New Jersey and elsewhere. The three storms together have created a heightened sense of risk, which is strengthened by new tools to assess future risks of sea level rise. Hurricane Sandy was an important event for water utilities, if not profoundly transformative. Some water utilities already had long-term plans in place to protect their infrastructure, and more are taking at least some action. Utility actions with the support of federal and state funds and regulatory guidance are mitigating future risks, and the storms at least briefly raised the profile of these essential services. The case study presented here indicates how a water utility company can take steps to improve ongoing operations in a manner that also improves its ability to respond to individual emergencies and to long-term threats. As with this example, utilities must constantly show customers and regulators that upfront investments provide demonstrable benefits, and wherever possible can actually reduce operational costs. The process can be seen as a form of "punctuated evolution" where new risk assessments result in a series of actions that lift utilities from a prior plateau to a new level. Only with new storms, updated risk assessments, and evolving policy to significantly modify the location and pattern of coastal and riverine development will water utilities be encouraged or forced through a further cycle of change.

LESSONS

To effectively manage risks to water utility systems in flood-prone areas, the federal government, states, local governments, and individual utilities will need to address several tasks.

- Utilities will benefit from risk assessments, creating detailed analyses of system components that may be harmed by storms and identifying the extent to which those components are critical to system functions (for example, pumps and treatment facilities).

Different utility components will be vulnerable to different stages and effects of climate change. Climate modelers project a number of conditions that water utilities will have to manage. In the short term, coastal water utilities must prepare for and respond to severe storms that may damage infrastructure. In the medium and long term, water utilities in coastal and estuarine areas must manage a variety of threats to their infrastructure in response to the rise of sea level. In the long run, water utilities throughout the region will likely face changes in water supply yields and in water supply and wastewater treatment needs, as rainfall and stream flow patterns change, reducing the effectiveness of some existing infrastructure. The federal and state governments will need to provide a framework, clear criteria, and incentives or mandates that provide justification and motivation for utility efforts.

- The creation of routine systems for asset management can ensure that the existing infrastructure is maintained and operated to maximize operational effectiveness and minimize lifecycle costs.

It is hard for a utility to win state or local government approval for raising utility rates to improve maintenance, but reducing the lifecycle cost of maintenance is one strategy that utilities can implement on their own. Efforts like those described in the case study can improve the physical management of the system while reducing maintenance costs.

- Utilities can mitigate risk by modifying critical utility components to significantly reduce the potential for system failure during a major storm event.

Utility assets in hazardous areas will be increasingly at risk of flooding, structural damage, and loss of customers as a result of building damages. Utilities can place some facilities and equipment in less risky locations, but they cannot avoid risks entirely and therefore must engage in efforts to reduce damages and enhance their resilience,[35] such as through the use of submersible pumps and creation of on-site generating capacity.

- Water utilities can play a role in changing the pattern and design of at-risk development so as to reduce physical impacts from major storm events.

As long as development exists in hazardous locations, water utility services must provide water supply, sewage collection, and stormwater management, because utilities have no authority to remove services from existing developed areas. Therefore, their fate is tied to that of at-risk development. Where development is removed through government buyouts or where buildings are destroyed and not replaced, the utility services will also be removed. Utilities have incentives to educate officials and residents about the ongoing costs that consumers and taxpayers are paying for development in hazardous zones. However, utilities have no regulatory authority and therefore can play only a limited role in this issue without backing from federal and state government guidance and regulations.

Successful action by water utilities and others to address the threats of severe events requires a recognition and quantification of the current and future risks involved, clear policies forcing action, and a willingness to invest prior to major events as a means of reducing damages.

NOTES

The author gratefully acknowledges the willingness of the New Jersey American Water Company to provide data and graphics information regarding the case study. Thanks go especially to Vincent Monaco, P.E., Asset Planning Manager, and Christopher Kahn, MGIS/GISP, Senior GIS Project Manager, both of the Engineering Division, NJ American.

1. Alyson Kenward, Daniel Yawitz, and Urooj Raja, "Sewage Overflows from Hurricane Sandy," Climate Central, 2013, http://www.climatecentral.org/pdfs/Sewage.pdf.
2. U.S. Environmental Protection Agency, *USEPA Region 2: Climate Adaptation Plan* (New York: EPA, 2013), http://epa.gov/climatechange/Downloads/impacts-adaptation/region-2-plan.pdf.
3. Kenward, Yawitz, and Raja, "Sewage Overflows."
4. County of Nassau, *Bay Park Sewage Treatment Plant Super Storm Sandy Recovery* (County of Nassau, NY: Department of Public Works, 2013), http://www.citizenscampaign.org/PDFs/Bay%20Park%20STP%20Sandy%20Status.pdf; Tom Brune, "FEMA to Give $80M More for Bay Park Sewage Plant," *Newsday*, January 17, 2014, http://www.newsday.com/long-island/nassau/fema-to-give-80m-more-for-bay-park-sewage-plant-1.6824693.
5. New York City, *A Stronger, More Resilient New York* (New York: New York City, 2013), 17, http://www.nyc.gov/html/sirr/html/report/report.shtml. New York City reported that 560 million gallons of partially or untreated sewage were released from treatment plants that either lost power or were damaged, less than one day of normal flow citywide.
6. Ibid., 74–76.
7. Billy Joel, "Allentown" (*The Nylon Curtain*, 1982). Allentown lyrics are available at BillyJoel.com, http://www.billyjoel.com/music/hits/allentown.
8. Climate Central has mapped relevant information for the New Jersey coastal areas, such as property values that in many places range from $10 million to more than $100 million per acre, often for areas with small homes. Similar information is available for New York City and Long Island. See http://sealevel.climatecentral.org.
9. The NJ Flood Mapper (http://www.NJFloodMapper.org) online tool is especially instructive in this regard, showing the areas that are currently flooded at mean higher high water (MHHW) and how that would change with increases in sea level at one-foot increments. Storm surges would thereby add to a higher base water level.
10. New York City, *A Stronger, More Resilient New York*, 6.
11. Ibid., 24.
12. Ibid., 30.
13. Ibid., 213.
14. Ibid., 50–56.
15. New Jersey Office of Emergency Management, *State of New Jersey 2014 State Hazard Mitigation Plan (draft)* (Trenton, NJ: NJOEM, 2014), http://www.state.nj.us/njoem/programs/mitigation_plan2014.html.

16. New Jersey Department of Community Affairs, *Post Sandy Planning Assistance Grant: Program Description and Guidelines* (Trenton, NJ: NJDCA, 2013), http://www.nj.gov/dca/services/lps/pspag.html.

17. See the website of the New Jersey Department of Environmental Protection (NJDEP), Division of Water Quality, http://www.nj.gov/dep/dwq.

18. The surveyed organizations were the American Water Works Association–NJ Chapter (a professional association regarding water supply), the Association for Environmental Authorities (for municipal utility authorities in water supply and sewage systems), the NJ Utilities Association (for investor-owned companies) and the NJ Water Environment Association (professional association regarding wastewater management).

19. Daniel J. Van Abs, *Stakeholder Engagement Report: Water Resources; Climate Change Preparedness in New Jersey* (New Brunswick, NJ: Rutgers University on behalf of the New Jersey Climate Adaptation Alliance, 2013). Available, along with other reports, at http://njadapt.rutgers.edu/resources/njcaa-reports.

20. Ibid., 10.

21. Ibid., 8.

22. Ibid.

23. The municipalities served by NJ American Water Company in this area are Bay Head Borough, Lavallette Borough, Mantoloking Borough, and the barrier island sections of Berkeley, Brick, and Toms River Townships. All are located on the northern section of Barnegat Bay Island.

24. All information regarding this project was provided through interviews of Vincent Monaco, P.E., Asset Planning Manager, and Christopher Kahn, Engineering Division, NJ American Water Company. Data and maps were prepared by Mr. Kahn and provided by NJ American, and are used with their permission.

25. U.S. Environmental Protection Agency, *Climate Ready Water Utilities: Adaptation Strategies Guide for Water Utilities* (Washington, DC: EPA, 2012), http://water.epa.gov/infrastructure/watersecurity/climate/index.cfm.

26. Hurricane Sandy Rebuilding Task Force (HSRTF), *The Hurricane Sandy Rebuilding Strategy* (Washington, DC: U.S. Department of Housing and Urban Development, 2013), 49, http://portal.hud.gov/hudportal/HUD?src=/sandyrebuilding.

27. Executive Office of the President, *The President's Climate Action Plan* (Washington, DC: Executive Office of the President, 2013), http://www.whitehouse.gov/sites/default/files/image/president27sclimateactionplan.pdf.

28. *Floodplain Management and Protection of Wetlands, U.S. Department of Housing and Urban Development* (Federal Register FR-5423, 2013).

29. New York City, *A Stronger, More Resilient New York*, 215–217.

30. NJDEP, *Clean Water Financing Proposed Priority System, Intended Use Plan, and Project Priority List for Federal Fiscal Year 2014 (Including the Proposed Intended Use Plan for Superstorm Sandy CWSRF Financing)* (Trenton, NJ: NJDEP, 2013), 5, http://www.nj.gov/dep/dwq/pdf/cwf_2014P_cwpl.pdf. This document supports the NJ Environmental Infrastructure Financing Program, which is a joint effort of NJDEP and the NJ Environmental Infrastructure Trust, and addresses Hurricane Sandy financing for both clean water and drinking water programs.

31. NJDEP, *Infrastructure Flood Protection Guidance and Best Practices* (Trenton, NJ: NJDEP, 2014). Available, along with the other three guidance documents, at http://www.nj.gov/dep/dwq/mface.htm.
32. NJDEP, *Flood Hazard Area Control Act Rules, N.J.A.C. 7:13. Adopted Emergency Amendments and Concurrent Proposed Amendments: N.J.A.C. 7:13–1.2, 3.2 through 3.6, 7.2, 8.7, 8.8, 9.2, 10.4, 11.5, 11.6 and Appendix 2* (Trenton, NJ: NJDEP, 2013). These rules require development to use the best available flood elevation information, including the FEMA Advisory Base Flood Elevations (ABFE) at the one hundred–year (1% probability) flood level plus one foot of freeboard, which will be superseded in time with final Flood Insurance Rate Maps (FIRMs). However, the ABFEs do not include data from the most recent large storms in New Jersey—Hurricanes Irene and Sandy and Tropical Storm Lee.
33. Claudia Copeland, *Hurricane-Damaged Drinking Water and Wastewater Facilities: Impacts, Needs, and Response* (Washington, DC: Congressional Research Service, 2005), http://www.hsdl.org/?view&did=463973.
34. Federal Emergency Management Agency, *Hurricane Ike Impact Report* (Washington, DC: FEMA, 2008), http://www.fema.gov/pdf/hazard/hurricane/2008/ike/impact_report.pdf.
35. HSRTF, *The Hurricane Sandy Rebuilding Strategy*, 36. Resilience is defined as "the ability to prepare for and adapt to changing conditions and withstand and recover rapidly from disruptions."

13 • IMPACT OF EXTREME EVENTS ON THE ELECTRIC POWER SECTOR

Challenges, Vulnerabilities, Institutional Responses, and Planning Implications from Hurricane Sandy

FRANK A. FELDER AND
SHANKAR CHANDRAMOWLI

Climate change and extreme weather events have serious implications for the electric power sector. Key changes in the climate that we witness today include a rise in global surface temperature, changes to hydrological cycles, ocean acidification, rise in mean sea levels, and perhaps higher incidences of extreme weather events like storms and surges, heavy precipitation, and heat waves.[1] Disruption to the generation and supply of electricity could be considerable because of these environmental changes, and an assessment of the impact of climatic changes on the electric power sector is of increasing importance to policymakers across the world. As the electric power infrastructures in coastal areas are the most vulnerable, and as the effects of Hurricane Sandy are still fresh in the minds of many power sector stakeholders, in this chapter we use the recent experience with Hurricane Sandy in New York and New Jersey to alert institutions and policymakers to the vulnerabilities of the energy sector to climate change and extreme events.

In what follows we take the case of Hurricane Sandy to assess the effects of the changing climate on the electric power sector to inform planning measures. These measures usually take the form of adaptation and mitigation strategies implemented by government and private institutions, for example, utilities, generators, and consumers. In the first section, we present an overview of the electric power system with special focus on the nature of the electric grid, including generation, transmission, and distribution, and the sector's institutional framework. In the second section, we outline the key challenges and vulnerabilities of the electricity sector. The third section examines the institutional and policy responses to Hurricane Sandy by public and private entities based in New York and New Jersey. The final section summarizes the planning and policy implications for dealing with the challenges and risks associated with climatic change and extreme events.

OVERVIEW OF ELECTRIC POWER SECTOR

Electricity is an integral part of the global energy system and is an indispensable element for overall societal welfare and economic growth. Since the introduction of electrical power as a commercial energy source in the 1890s, the core technical aspects of the electric power sector have changed surprisingly little. A typical electric power system is represented in figure 13.1. Electric power systems have centralized or distributed power plants that generate electricity, often located near an energy source, say a coal mine or a dam, to minimize fuel transportation costs. Many power plants are located away from heavily populated load-centers. To transport electricity economically over these long distances, the generated power is transmitted at a high voltage that is stepped-up at the power plant and then stepped-down at substations, using transformers at both ends. From the substations, the power is distributed to retail consumers, including residential, commercial, and industrial customers, which use power at different voltage levels, by means of distribution lines. Presently, the United States has three main grid networks that are largely independent of each other: Eastern Interconnect, Western

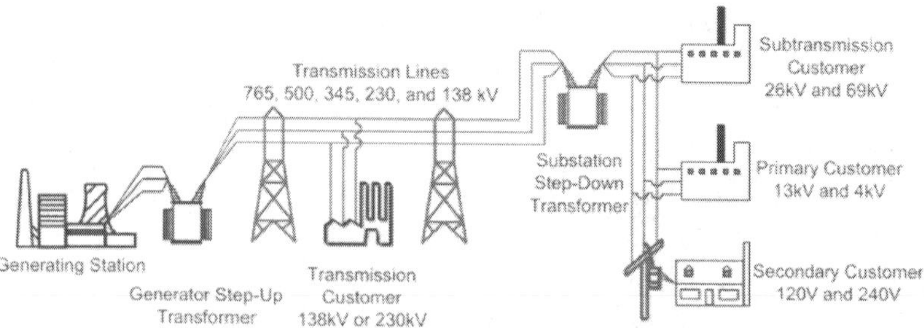

FIGURE 13.1. Key components of an electric power system. (Source: Nils J. Diaz, William J. S. Elliott, Linda J. Keen, Bob Liscouski, David Meyer, Thomas Rusnov, and Alison Silverstein, *US-Canada Power System Outage Taskforce Report—August 14th Blackout: Causes and Recommendations* [Washington, DC: U.S. Department of Energy, Office of Electric Transmission and Distribution, 2004], 5.)

Interconnection, and the Electric Reliability Council of Texas (ERCOT). New York and New Jersey are part of the Eastern Interconnect grid network.

A range of public and private institutions manage and regulate the electric power system infrastructure. Currently, there are about 2,200 traditional utilities in the United States handling the entire cycle, from power generation to power delivery, for more than 145 million customers of various consumer classes.[2] These traditional utilities can be investor-owned and publicly traded; publicly owned, usually at a municipal level; cooperatives in which customers own some portion of the grid; or federally owned. There are also about 1,700 power producers of the following types: traditional utilities, also called qualifying facilities, that support renewable energy and combined-heat-and-power units under the Public Utility Regulatory Policies Act of 1978 (PURPA); independent power producers that generate and sell electricity at wholesale market-rate to traditional utilities; and other stand-alone combined-heat-and-power units.[3]

Traditional electric power utilities in the United States are subject to federal, state, and local laws on aspects ranging from corporate practices to local environmental impacts. Except for Texas ERCOT, which is not operationally synchronized with the rest of the country, interstate transmission and the interstate sale of electricity on the wholesale market between utilities is regulated by the Federal Energy Regulatory Commission (FERC). FERC also regulates interstate transmission, issues guidelines for grid reliability, and oversees market and operational practices of system operators. In addition to FERC, the grid reliability at the regional generation and transmission levels is enforced by independent system operators (ISOs) or regional transmission organizations (RTOs), which are regulated, nonprofit entities. New Jersey is governed by an ISO called PJM Interconnection, which handles the grid operations and administers the competitive wholesale market in all or parts of thirteen states and the District of Columbia.[4] The New York ISO (NYISO) administers the wholesale markets and operates the grid for the state of New York alone (including Long Island).

At the state level, electric utilities are also subject to state public utilities commissions (PUCs), which regulate electric distribution companies, including retail electricity sales. State and federal environmental regulators also play a critical role in regulating all parts of the electricity supply chain. As our brief overview of the electric power system shows, reducing coastal threats to the electric power sector in the United States will require changes across a complex and multijurisdictional regulatory landscape.

Further complicating coastal adaptation is the nature of the electricity market and market regulations. The commercial market for electricity in the United States began as a franchisee-based monopolistic system but has shifted to a combination of competition-based systems of electric markets and regulated franchises. Until the early 1990s, the power sector in the United States, and much but not all of the rest of the world, operated on a rate-base paradigm of cost regulation in which the allowed rate of return was directly related to the cost of capital. For decades under these arrangements, the real average price of electricity showed a declining trend.[5] Under this scenario of declining average costs and high demand growth, the electric power industry's main concern was to meet demand by investing in new capacities. However, in the mid-1970s, the average retail prices began trending upward, mostly as a result of decreasing returns to scale at the generation side, moderation of demand growth, and, most important, higher fuel prices. Also, the utilities under a regulatory regime had the perverse incentive to increase profits by expanding their rate base. By expanding their rate base by means of excess capital investments, utilities were entitled to earn a rate of return from ratepayers in excess of the cost of capital. This tendency could lead electric utilities to adopt an excessively capital-intensive technology and consequently pass on the higher retail rates to final ratepayers.[6]

The PURPA Act of 1978 served as a catalyst for independent investors to enter the electric power sector. This also put pressure on generators to improve their production efficiencies. These underlying market changes, accompanied by an overall macroeconomic policy shift toward deregulation, led to the restructuring of electricity markets in the United States starting in the late 1980s.[7] It was felt that a restructured market could close the gap on some of the market inefficiencies in the production and pricing aspects of the power sector. The competitive framework hinges on the principle of equating the price of the good to the marginal costs of meeting the demand. For the electricity market, this translates as the marginal variable cost of the last generator unit that needs to be dispatched to meet the system demand.[8]

As of 2010, fifteen states, including New Jersey and New York, in the United States have adopted a restructured electricity market with different degrees of competition in the retail and wholesale markets.[9] Ten out of the fifteen states that have implemented market restructuring are based in the Northeast, partly in response to retail prices that are higher than the nationwide average. Seven states, including California, have tried to implement restructuring in the past but have now fully or partially suspended such efforts. The other states continue to operate under a traditional rate-base paradigm. Some of the key electricity statistics for the New Jersey and New York State region are summarized in figure 13.2.

KEY CHALLENGES AND VULNERABILITIES OF ELECTRIC POWER SYSTEMS TO CLIMATE CHANGE AND EXTREME EVENTS

Increasing temperatures, changes in freshwater availability, increased incidence of extreme storm events, and periodic coastal flooding are expected to affect the ability to generate, transmit, and distribute electrical energy to end-customers. The environmental implications of climate change are especially serious for the electric power sector because it consists of a centralized and interconnected network of generation, transmission, and distribution assets spread over a wide geographic footprint. A thorough assessment of the impact of climate change and extreme events on the electric power sector is necessary for implementing a forward-looking adaptation and mitigation plan. Any such plan should take into consideration the following anticipated climatological impacts.[10]

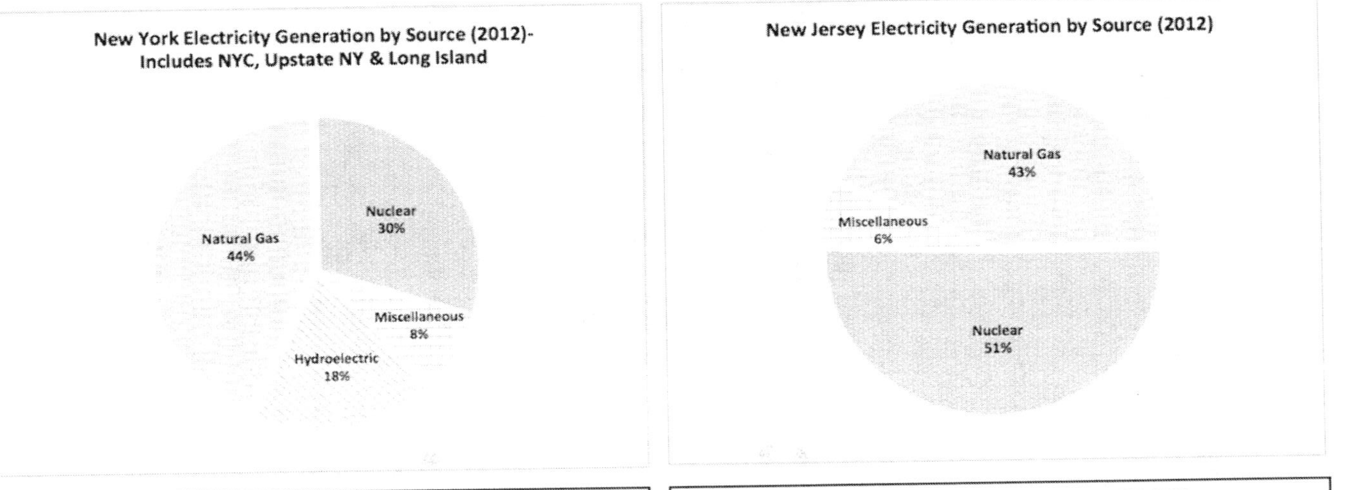

Total Demand for Electricity in New York State in 2012 is 143,163 GWh
Total installed capacity (as of 2010) is 39,357 MW (summer capacity)
Average retail price of electricity in New York (for 2012) is 15.15 cents/kWh Retail Prices by sector (for 2012): Residential- 17.62 c/kWh; Commercial- 15.06 c/kWh; Industrial- 6.70 c/kWh

Total Demand for Electricity in New Jersey in 2012 is 75, 053 GWh
Total installed capacity (as of 2010) is 18,424 MW (summer capacity)
Average retail price of electricity in New Jersey (for 2012) is 13.68 cents/kWh Retail Prices by sector (for 2012): Residential- 15.78 c/kWh; Commercial- 12.78 c/kWh; Industrial- 10.52 c/kWh

FIGURE 13.2. Comparison of key electricity statistics for New Jersey and New York. (Source: U.S. Energy Information Administration, Forms 860, 861, 923 for New Jersey and New York, http://www.eia.gov/electricity/data/state.)

Increase in Energy Demand

It is generally predicted that an increase in mean ambient temperature is likely to reduce the heating demand for colder regions in Europe and North America. This reduction is more or less offset by the increase in energy demand for cooling during the summer months. The overall energy demand is therefore expected to increase, owing to growth in population, economic output, and climate change.

Increase in Electricity Demand

A combination of warming weather and increasing adoption of air conditioners is expected to substantially increase the overall electricity demand, particularly during peak hours. The system peak demand for the region is expected to increase sharply in the coming decades, resulting in the addition of peaking and intermediate generating plants in the future (mostly combined cycle and combustion turbine units, which emit less pollution than coal power plants) unless substantial policy changes occur. Peaking units operate only a relatively few number of hours a year, that is, during hot summer days and emergencies; intermediate units operate 40 to 60 percent of the time, that is, during the workweek.

Increased Vulnerabilities of Power Plants

Nuclear-, coal-, and large natural gas–fired power plants are vulnerable as a result of decreased water availability and increased ambient air and water temperature. These plants require adequate supplies of freshwater for cooling purposes. As the temperature differential between the plant generator and the environment decreases, the net power generated also decreases. Although the drop in efficiency is typically small at a unit plant level, its cumulative supply-side impact could be large. Freshwater availability is expected to play a crucial role in facility siting decisions in the future.

Changes in the Hydrological Cycle

Changes in the hydrological cycle, including changes in river flow, increased precipitation and evaporation, melting of freshwater glaciers, and changes in the capacity of reservoirs, are expected to affect hydropower production. Increased precipitation and river flow in some river basins could yield a greater potential for hydroelectric generation, but only if the flow does not exceed the designed capacity of the reservoir. Hydropower is not a dominant source of electricity nationally, but is important in some states, including New York.

Changes in Demand for Renewable Energy Technology

Climate change is also expected to affect renewable energy technologies like wind and solar. Variations in demand may not match with the generation output from these units, resulting in inadequate or excessive backup capacity from fossil-fueled based units. To date there have been very few quantitative research studies on the impacts of potential climate change on solar energy production.

Increased Risk of Blackouts

Higher ambient air temperature reduces the prospects of radiative cooling from the transmission lines, which in turn can reduce the carrying capacity of the line. It can also lead to sagging power lines, which increase the risk of blackouts.

Increased Risk of Transformer Failures

The transformers at substations are also vulnerable to increased failure rates owing to temperature effects of climate change. Higher ambient temperature results in higher conductor temperature within the transformer, which in turn reduces the rated capacity of the transformer.

Increased Vulnerability of Electric Power Infrastructure

Electric power infrastructure located at coastal areas is at risk from increasing storm surges, flooding, and coastline erosion. Infrastructure located in riverine floodplains is likewise at risk from flooding.

To better understand the challenges in planning and responding to future extreme events, it is critical to analyze our responses to past events like Hurricane Sandy. Sandy was a post-tropical cyclone that hit the southern New Jersey coast on October 29–30, 2012.[11] At landfall of the storm eye, Sandy was officially designated as a Category 1 post-tropical storm with sustained wind speeds up to 80 miles per hour (or 70 knots), though it lost hurricane status only just before landfall. The storm surge was the highest recorded on the New Jersey, New York City, and Connecticut coastlines, ranging from three feet to nine feet above mean low tide level. The coastal inundations were also widespread in this region. As a result of this tropical storm, nearly

650,000 homes were destroyed and about 8.2 million retail customers lost power in the northeastern United States. Nearly 5 million of these customers were located in the New Jersey–New York City region.

Figure 13.3 presents a time line of power outages as Sandy struck the region and the immediately following period of restoration. Ten days after the storm made landfall, nearly seven hundred thousand customers in the Northeast were still without power.[12] The damage to electric power infrastructure in the region was unprecedented. On the generation side, directly after the storm nearly 8,000 megawatts of generating capacity in New York State was offline owing to flooding or damages. Ten days after the storm, nearly 2,200 megawatts of generation was still unavailable for dispatch in New York State.[13] Wind damaged above-ground power lines throughout the region, and storm surges in the New York City area destroyed parts of the underground network of distribution lines. Flooding of basements resulted in the loss of backup generator power for some hospitals and offices in lower Manhattan and other parts of the city.[14]

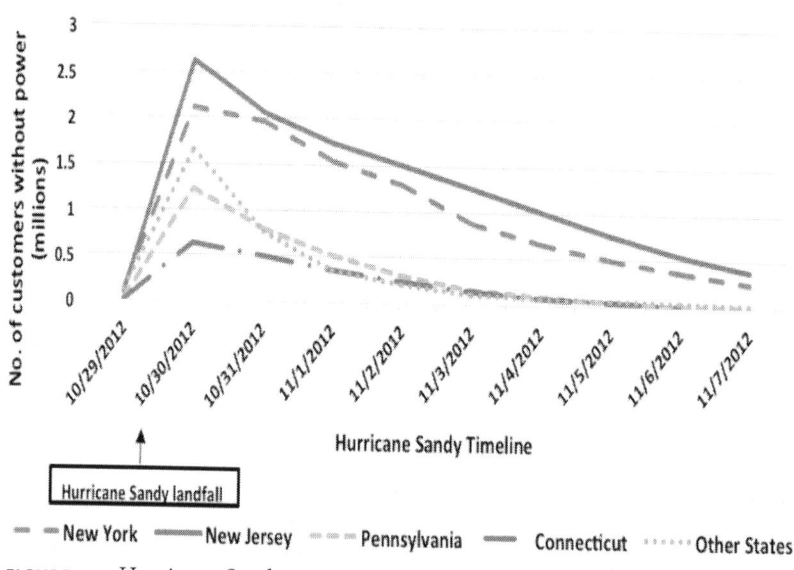

FIGURE 13.3. Hurricane Sandy power outages and restoration time line. (Source: U.S. Department of Energy, Office of Electricity Delivery and Energy Reliability, *Emergency Situation Reports—Hurricane Sandy*, http://www.oe.netl.doe.gov/named_event.aspx?ID=67.)

One key challenge observed during the storm was the interdependency of the oil and gas sector, the water and wastewater sector (as described by Daniel Van Abs in chapter 12), and other essential services with the electric power sector. What captured the most attention after Sandy was its major disruption to refining and pipeline distribution operations on the coasts of New York and New Jersey. There were also widespread gasoline shortages because of power outages at retail gas stations. Thus a combination of damages to ports and refineries, widespread power outages, and high demand for gasoline fuel led to gas-rationing policies in New York and New Jersey in the immediate aftermath of the storm.

Utilities are likely to incur huge labor and material costs to restore their systems from these and other damages, to be recovered from ratepayers in the coming years. A rough estimate of the cost incurred by the electric power sector is likely to touch $2 billion for the utilities based in the New Jersey-New York City region (see table 13.1). These estimates do not include the costs incurred by residential customers, commercial businesses, and industries owing to power outages.

TABLE 13.1 Estimated Cost of Repair and Restoration of Utility Assets in NJ–NYC Region

Utility (state)	Estimated cost of recovery
Jersey Central Power and Light (JCP&L; a subsidiary of FirstEnergy, Inc.) (New Jersey—north and south)	$680 million
Public Service Electric and Gas (PSE&G) (New Jersey—central region)	$420 million
Consolidated Edison (Con Ed) (New York City)	$521 million
Long Island Power Authority (LIPA) (New York—Long Island)	$800 million

SOURCES: Long Island Power Authority, November 19, 2012, http://www.lipower.org/newscenter/pr/2012/111912-update.html; "Superstorm Sandy Was Super Expensive for Con Edison," Forbes.com, May 9, 2013, http://www.forbes.com/sites/williampentland/2013/05/09/superstorm-sandy-was-super-expensive-for-con-edison; Jersey Central Power and Light Submits Rate Filing for Hurricane Sandy Restoration Costs, FirstEnergy.com, February 22, 2013, https://www.firstenergycorp.com/content/fecorp/newsroom/news_releases/jersey-central-power---light-submits-rate-filing-for-hurricane-s.html.
NOTE: Cost estimates are based on publically available press releases and articles. The cost estimates also include the repair and restoration of natural gas services in the region.

RESPONSES TO HURRICANE SANDY BY STAKEHOLDERS INVOLVED IN THE ELECTRIC POWER SECTOR

This section summarizes the short-term responses to Sandy from important stakeholders like utilities, regulators, and the federal and state governments. The federal government, specifically the Federal Emergency Management Agency (FEMA), deployed "assessment teams" on the ground at various states in the Northeast to monitor and address the problems caused by the storm. The Obama administration waived the requirement of proportional state government funding for relief efforts, thereby increasing the federal component of relief money substantially. The U.S. Department of Energy also engaged in daily coordination calls with operational heads or chief executive officers of utility companies to assess their needs and to prioritize restoration efforts. On an unprecedented scale, nearly sixty-seven thousand utility personnel from more than eighty electric utilities across United States participated in the immediate repair and relief effort under a program of mutual assistance.[15] Within two weeks after Sandy's landfall, power was restored to nearly 99 percent of the customers who lost power. The DOE also published periodic situation reports to update the status of repair and restoration services (see source in figure 13.3).

Apart from repair and restoration efforts, the utilities and regulators have also embarked on various hardening measures to improve the resiliency of the electric power sector. The hardening measures can be roughly classified under two types: tangible and soft types. Vegetation management, such as trimming tree branches that overhang or grow near power lines is a low-cost tangible type of remedy. PSE&G, a utility based in New Jersey, had to remove or trim nearly forty-eight thousand trees in order to restore power.[16] Other longer-term tangible hardening measures involve increasing the grade height of substations, transformers, and other key installations to protect them from storm surges. They also include using advanced metering infrastructure to track power outages in order to facilitate power restoration and improve communications with end-customers, and supervisory-control and data-acquisition systems that can control vulnerable distribution equipment from a central location, thereby minimizing damages to the electric power grid and speeding the restoration of power. Soft responses to improve energy sector resiliency include better emergency coordination in repair and restoration work, more resilient information and communication infrastructure, better emergency response planning, and improved coordination with out-of-state utilities.

LESSONS

There is strong evidence to suggest that future extreme weather events will compromise large swaths of the electric power sector, cause significant disruptions to daily lives, and inflict tremendous economic damage. Many of the analyses in this book observe how, in reaction to Sandy, local, regional, and national actors were motivated to make short-term policy and operational shifts to address crisis situations. Other analyses, like that in chapter 2 by Daniel Hess and Brian Conley, speak to how the lessons learned from prior storm events like Hurricane Katrina motivated emergency managers in Connecticut to plan ahead for transformational change on the path toward resilience, which better prepared them to manage the response to Hurricane Sandy. As our examination of the impacts of Sandy on the electric power sector in New Jersey and New York makes clear, advance planning with a multipronged approach is required to develop a more resilient power sector with capabilities to handle climatic shocks. In this section, we outline specific key planning and policy lessons for power sector stakeholders.[17]

- Investing in grid-hardening measures can considerably improve the power grid's capacity to withstand climatic shocks.

Utilities should prioritize investments based on cost effectiveness when evaluating measures like vegetation management, distribution and transmission line strengthening, and the protection of power assets from storm surges and flooding.

- Investing in emerging information and communications infrastructure can improve responses to power outages.

Utilities and system operators should focus on deploying cyber infrastructure technologies like advanced supervisory control and data acquisition, advanced metering infrastructure, outage management systems, or distribution management systems.[18] These technologies contribute to a better situational awareness of the grid and a more efficient response to power outages.

- Regulators must factor the costs of future hazard events into analysis of the costs and benefits of grid modernization investments.

State public utilities commissions and system operators must consider the long-term societal benefits of hardening and modernization investments incurred by utilities. These investments are likely to be recovered from ratepayers in the form of a socialized rate-base. Any cost-benefit analysis by the regulators must also take into account the fact that the incidence of extreme events is likely to increase in the coming years and that such shocks can inflict enormous costs on society.

- Developing weather and climate projections and hazard prediction models can help to accurately identify key vulnerabilities.

Electric utilities and system operators must closely work with the climate science community to develop accurate short-term weather projection models and perhaps long-term climate models, with emphasis on predicting the potential damages to electric power assets. This would translate into incorporating weather-related phenomena like hurricanes, storm surges, and flooding patterns into impact assessment and vulnerability models for the electric power sector.

- Standardizing procedures for mutual assistance programs for utilities can improve response time in repairing the grid.

Mutual assistance responses are vital in responding to widespread disruptions caused by climatic shocks. Electric utilities across the country must continue to collaborate to improve their abilities to work together by integrating out-of-state personnel and resources in grid repair and restoration works.

- Coordinating the emergency response planning process across different tiers of stakeholders can improve the efficiency and timeliness of repair and restoration.

The scale of Sandy's destruction necessitated direct coordination between utility company chief executive officers or operational heads with the federal government. This communication helped to coordinate the repair and restoration efforts across the region and also provided situational awareness of the scale of destruction to the federal government. There is a need to standardize these vertical communication channels and develop communication protocols to better handle the emergency response.

- Developing appropriate oil and gas sector response plans will enhance the effectiveness of electric power sector responsiveness.

Sandy exposed the interdependencies between the oil and gas sector and the electric power sector. It also highlighted the need to implement a coordinated response plan on the regional and national scales for the oil and gas sector by the federal and state governments. The oil and gas sector could also explore the possibility of establishing a mutual assistance program along the lines of the existing initiative in the electric power sector.

- Acknowledging the trade-off between investing in mitigation and adaptation efforts will better position policy actors and others to make the best choices on the path toward future resilience.

Finally, utilities, regulators, and ratepayers must also acknowledge trade-offs between investing in mitigation efforts versus adaptation efforts. Mitigation efforts like distributed renewable generation, microgrids, and energy efficiency would reduce climate change damages by reducing carbon emissions and therefore reduce the vulnerability of the system to extreme events.

Extreme storm events, like Sandy, have highlighted the infrastructural vulnerabilities and also the institutional shortcomings in responding to such events. The various grid-modernization technologies and hardening measures identified here have the potential to reduce power outages in the future. Likewise, if implemented, the institutional reforms outlined in this chapter can ensure a faster recovery from climate-related shocks. Utilities, regulators, federal/state governments, and ratepayers must work collaboratively toward implementing these recommendations for a more resilient power sector in the future.

NOTES

1. Intergovernmental Panel on Climate Change, *Climate Change 2013: The Physical Science Basis,* contribution of Working Group I to the Fifth Assessment Report of the IPCC (Cambridge: Cambridge University Press, 2013), "Summary for Policymakers," 5–8.
2. U.S. Energy Information Administration (US EIA), Form 861 (2012), http://www.eia.gov/electricity/data/eia861.
3. Combined heat and power units coproduce heat and electricity. They primarily generate heat (i.e., steam) for captive industrial or commercial processes. Electricity generated from surplus heat is used for self-consumption or for sale to utilities.
4. The abbreviation *PJM* reflects the original three states in its formation: Pennsylvania, New Jersey, and Maryland.
5. Edison Electric Institute, *Assessing Rate Trends of US Utilities* (Washington, DC: Edison Electric Institute, 2006).
6. This is called *gold-plating* and is also referred to as the Averch-Johnson effect in regulatory parlance. See Harvey Averch and Leland L. Johnson, "Behavior of a Firm under Regulatory Constraint," *American Economic Review* (1962): 1,052–1,069. Also see Alfred Kahn, *Economics of Regulation* (Cambridge, MA: MIT Press, 1988), on the regulatory trends of public utlities in the United States.
7. Paul L. Joskow and Richard Schamalensee, *Markets for Power: An Analysis of Electric Utility Regulation* (Cambridge, MA: MIT Press, 1983); Kahn, *Economics of Regulation.*
8. William Hogan, *Competitive Electricity Market Design: A Wholesale Primer* (Cambridge, MA: Harvard University Press, 1998).
9. US EIA, http://www.eia.gov/electricity/policies/restructuring/restructure_elect.html.
10. For a more detailed account of the reviewed studies, readers are encouraged to refer to our earlier paper: Shankar Chandramowli and Frank Felder, "Impact of Climate Change on Electricity Systems and Markets: A Review of Models and Forecasts," *Sustainable Energy Technologies and Assessments* 5 (2014): 62–74. Also see Department of Energy (DOE), *US Energy Sector Vulnerabilities to Climate Change and Extreme Weather* (Washington, DC: U.S. Department of Energy, 2013).
11. This synoptic summary is primarily based on a tropical cyclone report prepared by the National Oceanic and Atmospheric Administration, written by Eric S. Blake, Todd B. Kimberlain, Robert J. Berg, et al., *Tropical Cyclone Report: Hurricane Sandy* (Miami: National Hurricane Center–NOAA, 2013).
12. US EIA, http://www.eia.gov/todayinenergy/detail.cfm?id=8730.
13. NYISO, *Hurricane Sandy: A Report from New York Independent System Operator,* http://www.nysrc.org/pdf/MeetingMaterial/RCMSMeetingMaterial/RCMS%20Agenda%20158/Sandy%20Report%202-22-13%20Public%20Version%202.pdf.
14. Alexandra Sifferlin, "Lessons from Storm Sandy: When Hospital Generators Fail," *Time,* October 30, 2012, http://healthland.time.com/2012/10/30/lessons-from-storm-sandy-when-hospital-generators-fail.
15. The response of the federal government is documented in a Department of Energy report, "Overview of Response to Hurricane Sandy: Nor'easter and Recommendations

for Improvement," February 26, 2013, http://energy.gov/sites/prod/files/2013/05/f0/DOE_Overview_Response-Sandy-Noreaster_Final.pdf.

16. Todd B. Bates, "Superstorm Sandy Statistics for N.J., N.Y., etc.," *EnviroGuy* (blog), posted August 20, 2013, APP.com (*Asbury Park Press*), http://blogs.app.com/enviroguy/2013/08/20/superstorm-sandy-statistics-for-n-j-n-y-etc.

17. DOE, "Overview of Response to Hurricane Sandy." Readers are also encouraged to refer this DOE report for a detailed list of recommendations for the power sector. See note 15, http://energy.gov/sites/prod/files/2013/05/f0/DOE_Overview_Response-Sandy-Noreaster_Final.pdf.

18. For a description of these technologies, see the glossary section at Smartgrid, https://smartgrid.gov.

CONCLUSION

Emerging Responses to Life on the Urbanized Coast after Hurricane Sandy

DANIEL J. VAN ABS AND KAREN M. O'NEILL

Knowing that decisions made soon after a disaster can remake a community or region, we set out to discover how people responded during and after Hurricane Sandy, primarily through public and private institutions. Authors of chapters in this book presented case studies of people making decisions under particular circumstances that highlight the sorts of questions that all people in urbanized coasts are facing.

We described each chapter's main findings in the book's introduction, and so in this concluding chapter we draw out findings across the chapters to answer the central question for this book, whether Hurricane Sandy is a transformational event, just another disaster, or something in between. People who are transforming their attitudes or behavior may be doing so because they think Sandy was linked to climate change or more simply because they were shocked by the damage that one storm could cause. Final judgments on this question will come only after a decade or more. We have documented early responses that will affect the ultimate answer.

Three categories emerged. First are individuals and organizations that focused on recovering basic functions within months of the disaster, especially homeowners and resort and tourism operators, whose long-term plans rarely extend beyond reestablishing residency or business, short-term cash-flow needs, or at most, a twenty-year mortgage. Second are individuals and organizations that were already alerted to potential hazards and incorporated lessons into their ongoing processes for planning and management. Although the organizations in this category had immediate responsibilities, such as providing emergency services or restoring water service, they were also already committed to long-term planning for contingencies of all kinds. Sandy affirmed that their planning is important, and it gave them new information and often new resources. Third, for a few individuals and organizations, the storm was indeed a focusing event, raising concern or prompting new efforts to manage coastal hazards. This category is small and highly diverse, including residents who suffered damage and said they would leave the shore, survey respondents who viewed Sandy as a sign of climate change and therefore likely to be repeated, and a few officials and managers who sought to change their operations in some major fashion.

The exceptions to these three generalizations are worth watching, for identifying how intentions and efforts to reduce coastal hazards might spread beyond those who are already convinced that adaptation is important. For example, regarding the first category of people facing short-term pressures, which seem especially difficult to change, some town officials are starting to suspect that some kinds of land use are too costly over the long term, advocating steps such as the purchase and removal of homes in entire neighborhoods to reduce future risks. While recent buyouts in New Jersey targeted neighborhoods along tidal rivers, rather than beach areas, the success of these removal projects provides a proof of concept that could later be expanded on the coast. New York State immediately set up a buyout program for beach areas, although it has been limited to a few locations.[1]

Regarding the second category, for sectors where leading organizations do systematic long-term planning, anger about the poor performance of lagging organizations in a sector could increase customer and government pressure to improve their planning. Regarding the third category, of people who saw Sandy as a defining event, some who suffered overwhelming damage may experience it as an isolating experience, but others may become agents in persuading others to care about adaptation. The authors have also identified constraints on our abilities to recognize hazards and to spend money today to save property and lives tomorrow. Most of these constraints are familiar to hazard experts, if not to others. As an unusual storm, though,

Hurricane Sandy revealed the constraint of complacency, especially among people living in places that are not usually directly hit by hurricanes.

Therefore, the answer to our question of whether the storm transformed concerns about coastal vulnerability is that it harmed many, reinforced concerns for some, and transformed a few. The people who experienced Hurricane Sandy as a transformational event are few and varied because there has been little leadership in government and other major institutions to frame it as a transformational event. Such a framing could define storm risk or its exacerbation from climate adaptation and propose a set of policies, behaviors, and physical projects. Experts have provided concepts for such framing, but these expert voices are diffused. Without support from large-scale institutions for lending, planning, and regulating, even the people who became concerned because of Sandy will have difficulty finding the ideas, tools, encouragement, and policy framework to take action.

Two broad potential strategies for action that have failed to attract institutional attention in response to Sandy are reductions in emissions of climate-change gases (mitigation) and shifting population away from the most hazardous areas (adaptation). Adaptive strategies have not been prominently linked to climate-change mitigation, and adaptations often involved building more barriers to protect existing development or even to encourage more development. In places with high value land, investors find it a good bet to intensify land use in hazard zones. They expect that in the short- and medium-term a given site is unlikely to experience a direct hit from a storm and that governments will continue to subsidize this development through various policies. New York City may have created an ambitious plan for storm resilience, but it also continues to promote new coastal development, as Kenneth Gould and Tammy Lewis describe in chapter 7.

In the following sections, we discuss several supporting questions we had to answer in order to address our central question about transformation: Did Sandy change how people perceive risk, prompt them to act in significantly different ways, reveal constraints on adaptation, provoke cultural and emotional responses, or change what people think should be done in the long run? We end with likely scenarios for future life along the coast.

CHANGED EXPECTATIONS ABOUT RISK AND CHANGED AIMS

All of the case studies asked whether organizations and individuals changed their aims and practices because they had changed their expectations about risk and perceptions of coastal vulnerability. Psychological researchers have debated how perceptions are linked to behavior, but studies show that perceptions of risk are important because they can lead to protective behavior.[2] We found that only some organizations and politicians have responded so far to concerns that many members of the public are increasingly expressing about coastal risk.

Organizations that deliver risk messages are especially likely to review their performance after a disaster, sometimes changing their practices after learning how their audiences understand risk. Chapters about weather forecasters and local emergency managers show examples of this kind of reassessment. Daniel Hess and Brian Conley (chapter 2) report that because residents' experience of storms strongly influences whether they will evacuate for a newly predicted storm, emergency managers were concerned to communicate that Sandy was an unusual storm. Crafting such messages will be even more important if the patterns of future storms defy past experience, as discussed by Steven Decker and David Robinson (chapter 1) regarding the National Weather Service. Two other sectors that are directly vulnerable to storms, fisheries and utilities, similarly took information about risk learned after Sandy and incorporated this into their ongoing activities or planning.

By contrast, several chapters make clear that politicians and leaders of the tourism and real estate industries throughout the Sandy region avoided discussing the storm as an indicator of risk that should lead to changes in coastal development. Leaders in New Providence, the main tourist island in the Bahamas with little history of hurricanes, had a moment of concern after Sandy, but soon returned their focus to attracting tourists, as Adelle Thomas discusses in chapter 3. Mark Hewitt (chapter 8) and Briavel Holcomb (chapter 9) explain that the impulse to dismiss risk and to rebuild quickly after Hurricane Sandy repeats a long historical pattern at the Long Island and New Jersey Shore areas. This pattern is being repeated despite the long-term fiscal costs of the "déjà vu" development pattern, as discussed by Clinton Andrews (chapter 10). Gould and Lewis explain that developers and officials in New York City have sidestepped discussing the risks of Brooklyn's Gowanus Canal area, including its legacy as a Superfund site and its inundation by Sandy.

The general populace, though, did want to talk about risk. Public opinion polls analyzed by Ashley Koning and David Redlawsk (chapter 4) show that a majority of residents were already concerned about a changing climate and that Sandy heightened concerns about coastal vulnerability. People felt the storm transformed lives, landscapes, and attitudes, viewing it as a "wake up call."

PLANS ALREADY UNDER WAY

The supporting question that would most clearly indicate if Hurricane Sandy was transformational is whether people have already made changes in actions or near-term plans. We found that the storm has inspired a few efforts to change human and natural systems to adapt to the effects of both current storm risk and climate change. Some individuals, organizations, and governments may wish to take action but lack the resources to do so. This section looks at who took action based on changed expectations soon after the storm, and who did not.

The category of organizations that were already committed to long-term planning responded to Sandy as an event highlighting their needs to ensure reliable and responsive operations. Some of these organizations are responsible for communicating about hazards or responding to disasters, while others provide continuous services. In a way, Sandy was vindication of their past preparations.

Communicators and responders are represented in chapters on weather forecasters and emergency managers. As Decker and Robinson explain, the U.S. National Weather Service decided after Hurricane Sandy that allowing its high-profile National Hurricane Center to continue tracking storms once they reached the country's mainland would help signal that these storms could still be powerful, to avoid false perceptions about risks when a tropical system becomes non-tropical. It started this practice within months of Sandy. Essentially, it is moving from a scientifically correct but confusing approach to one that may lack technical rigor but is more likely to be understood by the public. Hess and Conley explain that local emergency managers in Connecticut felt their evacuation efforts for Hurricane Sandy were successful because they had reviewed their procedures after earlier disasters and drills. They also took up a new round of reviews after Sandy to improve their next responses. Conversely, managers in Cuba learned that their generally well-respected system for hurricane preparedness had essentially ignored an entire segment of the island nation, leaving its eastern coast vulnerable to extensive damages, as Thomas discusses. Clearly, even those known for looking ahead can do so with clouded vision.

Organizations concerned about reliability are represented in chapters on the water and electric sectors. Some utility companies are applying short- and long-term lessons from Sandy to their ongoing management plans and capital projects. The chapters on water services by Daniel Van Abs (chapter 12) and on electricity services by Frank Felder and Shankar Chandramowli (chapter 13) explain that some organizations in these sectors have taken action in response to Sandy, although portions of each of these sectors are lagging. In some situations, actions taken for entirely different reasons, such as to enhance customer service, provided serendipitous benefits in damaged areas. Both sectors are under contradictory pressures from regulators. Some regulations encourage adaptation, such as recovery grants from the Federal Emergency Management Administration and resilience requirements, while other regulations seemingly discourage adaptation, such as rate regulators' resistance to proposals to increase customers' utility bills to help fund infrastructure retrofitting.

Commercial fishermen did indeed respond, but the response is not properly characterized as a plan made in response to a storm, because they responded in much the same way they do to other events that threaten their livelihood. Fishermen interviewed by Angela Oberg and colleagues (chapter 6) already view themselves as having to adapt to changing conditions, although small operators do not typically project their plans far into the future. Commercial fishermen work so closely to physical and financial peril in the course of an ordinary fishing season that they experienced Sandy as within the range of what might be expected. The nimbleness of these small operators stands in contrast to large institutions with extensive infrastructure and facilities.

As noted, at the other extreme were most people involved in land development and tourism, whose efforts to patch things together to pre-storm conditions (using a "Restore the Shore" motto in New Jersey) were often their first and last responses to the storm. Chapters on municipalities describe how owners of business and residential properties with deep pockets, or who got federal aid or insurance payouts, generally rebuilt as quickly as they could on the same lots. They usually had to adhere to more stringent federal requirements to reduce flood hazards, but they rebuilt at the same vulnerable sites, with little resistance from local officials. Holcomb details how tourist areas in New Jersey and Long Island mounted the summer season of 2013, rebuilding boardwalks and other public facilities along the same oceanfront, even though many adjacent residential and commercial buildings could not be restored in time for the season. The boardwalks in some beach towns became a facade behind which devastated communities were hidden. Thomas describes an exception, where residents and officials in the outer islands of the Bahamas, called the Family Islands, improved emergency plans and efforts for adapting their infrastructure and moving residences away from the shore. These islands are poorer and less populated than the Bahamas' main tourist island and did not face as much pressure to rebuild quickly to sustain major flows of tourist revenue.

Sandy was clearly a signal event for wildlife managers faced with sand-stripped beaches that had formerly provided essential habitats for animals. Chapter 5, by Joanna Burger and Larry Niles, reports on managers' concerns that the storm would disrupt the migration of shorebirds, particularly the threatened red knots, migratory birds that feed on eggs laid by horseshoe crabs along a few key Delaware Bay beaches. Wildlife managers and nongovernmental environmental organizations formed an ad hoc group to replenish sand in time for the crabs' arrival. This event has prompted wildlife managers to begin planning so that they might provide for such emergency responses actions if needed in the future. However, funding constraints remain, as is true for many aspects of planning, response, and restoration programs.

FORCING FACTORS OR CONSTRAINTS

Even those individuals and organizations that wish to reduce their exposure to coastal hazards are limited in what they can easily achieve, so we asked what forcing factors or constraints they face. The most obvious constraint is that some activities dependent on the ocean, like fishing, require that people expose themselves to danger. But with few of us relying on direct access to the sea for our survival, most of the constraints today are human-made. Understanding the limitations posed by our existing physical infrastructure, institutions, social systems, and finances is important for creating policies that encourage people to make their own changes. Social values and many current policies perversely encourage settlement in hazardous sites at the shore and back bays. The existing transportation, energy, water, and public facility infrastructure systems represent billions of dollars in past expenditures that are not easily abandoned or modified. The organizations and individuals already committed to long-term planning usually do so in response to incentives or requirements from regulations, insurance costs, long-term loans or bonds, or customer demands. People involved in land development and tourism faced constraints that pushed them to a shorter-term orientation.

The need to capture seasonal income streams in an industry prone to small margins and frequent weather-based disruptions is the fundamental constraint of the tourism industry and of resource-poor coastal communities that are similarly dependent on tourism. This was clear in efforts in the United States and the Bahamas to restart the machinery of the tourist trade as soon as possible after Hurricane Sandy.

The economic need to increase the value of land through development is a forcing factor for a complicated array of business and political arrangements that intensify development along and near the waterfront in tourist areas and more broadly. Gould and Lewis illustrate through the case of the Gowanus Canal in Brooklyn the power of a thriving real estate market and municipal policies to remake polluted waterfront sites within storm surge zones into exciting new development opportunities. In small towns, residential properties often provide the bulk of property taxes, and officials are concerned to sustain those revenues by taking actions that increase the value of those properties. Andrews concludes, contrary to common political expectations, that continuation of current development patterns in towns with flood-prone areas will in the long run increase property tax rates, indicating the fiscal problems these hazardous zones present for such towns. Upcoming increases in the U.S. National Flood Insurance Program premiums may force a shift in land ownership to wealthier owners, displacing those who cannot afford to pay for flood insurance premiums or to rebuild with features that reduce those premiums.

The political need to act quickly, or at least appear to, after a major policy failure or disaster is a fundamental forcing factor for officials. Koning and Redlawsk show that linking his governorship closely to storm recovery raised New Jersey Governor Chris Christie's poll numbers for some months above pre-storm levels. But it ultimately created new political vulnerabilities for him as complaints accumulated that hurricane aid was coming too slowly and as allegations surfaced that his administration had sent aid funds to relatively undamaged towns in exchange for election endorsements.[3]

The political need to direct most or all public spending toward human constituents is the constraint on environmental and wildlife policies that are perceived as primarily yielding benefits to non-humans. Burger and Niles explain that although the federal government (with support of the states and municipalities) regularly replenishes beaches with sand to support recreational uses, there is no similar provision for restoring sand for animal habitats.

A variety of other constraining conditions that hazard experts routinely cite as concerns were also evident in the case studies. For example, vulnerable groups, such as socially isolated elderly, poor, or rural residents, tend to lack the ability to reduce or manage their exposure to hazards. Mariana Leckner and colleagues discuss differences in coping abilities of residents in New Jersey (chapter 11), and Thomas discusses the extreme exposure to risk for residents in rural areas in Haiti, who had already been displaced by a major earthquake. Andrews details federal, state, and local policies in the United States that allow or encourage settlement in hazardous areas and notes reforms proposed to reduce these perverse incentives. The costs of restoring infrastructure systems are noted in the chapters on electric and water services by Felder and Chandramowli and Van Abs. As long as buildings are retained or restored in a particular locale, utilities must be provided regardless of the risks to their infrastructure. The increased costs of supporting infrastructure in hazardous zones are spread across all ratepayers, just as the costs of subsidizing flood insurance, providing disaster relief, and rebuilding public facilities are paid by all taxpayers.

CULTURAL AND EMOTIONAL RESPONSES

Because people invest meaning in the places where they live, we asked whether Hurricane Sandy has changed how people think and feel about life along the coast. Several authors noted that politicians in the United States and the Caribbean were able to sense and express residents' alarm about damage and deaths soon after the storm, although politicians' abilities or interest in reflecting this public sentiment waned as the inevitable problems of recovery unfolded. Residents who lost relatives or friends or who had not received aid quickly had the strongest feelings of frustration, anger, or impotence, as described in the study of three towns by Leckner and colleagues and in opinion polls reported by Koning and Redlawsk. People already involved in long-term planning and hazard management institutions expressed the least dismay, even though the storm defied some of their expectations. None of the chapters identified people who feared that their way of life could disappear or be radically changed.

For the commercial fishermen interviewed by Oberg and colleagues, Hurricane Sandy was just another storm. This stoic reaction cuts across two types of fishing subcultures. Along the Jersey Shore and Long Island, these researchers found some fishermen who saw their work as a business, "a floating store," as one fisherman put it, versus others who saw it as a way of life. Investigating this latter subculture, their chapter noted that the cohesive local culture of a single fishery cooperative provided a focus for a shared sense of risk and sense of purpose and history that enabled members to interpret the storm as something they would survive.

Thomas explains that the storm was one in a series of disasters in Haiti that increased the disruptions experienced by four hundred thousand people who had lost their homes as a result of the 2010 earthquake and hurricane and were living in displacement camps, exacerbated by two hundred thousand additional citizens who evacuated to camps after Sandy. By contrast, while people in the Bahamas initially expressed concern that they were more vulnerable than they had assumed, official rebuilding efforts did not encourage a lasting public discussion of such concerns.

Hewitt explains that the idea of the shore as a getaway location continues to dominate thinking about the Sandy region in the United States, even though large areas of the shore now host year-round residents, diverse economies, and a rich and underappreciated network of ecological functions. Yet poll findings from Koning and Redlawsk indicate there is new unease about coastal hazards following Hurricane Sandy and concern about storm victims. Whether that general unease translates into specific actions by those who own land, run businesses, or visit the shore is a major question.

IDEAS ABOUT HOW WE SHOULD REACT IN THE LONG RUN

We also asked whether organizations and individuals expressed ideas about how they or others should react to coastal hazards over the long run. Most of the people who mentioned long-term prospects were concerned about these hazards even before the storm.

In polling reported by Koning and Redlawsk, a majority of New Jersey residents, most of whom were not shore residents, hoped officials would delay rebuilding until they had good assessments of long-term risks. Some of the residents who faced the immediate tasks of dealing with their own damaged houses, though, found it more difficult to think about the future, as reported in the chapter by Leckner and colleagues. Local government officials in the three towns they studied expressed more concern about the long-term future than their residents did, but officials varied in their capabilities to do anything about it. None of the three towns was considering major proposals for moving residents and businesses out of the most hazardous zones, although one town has decided to move municipal facilities out of a floodplain.

Institutions that rely on long-term planning as part of their ordinary work are already structured to incorporate the experiences of storms like Sandy into plans and treat their short- and long-term operations as interrelated. Van Abs explains how information technology being tested to improve routine operations aided a water utility's recovery after Sandy and may help utilities restore services at a reasonable cost in neighborhoods that are repeatedly devastated by storms. Felder and Chandramowli describe the sort of planning that any complex institution would ideally be developing in order to reduce harms from climate change. Their research presents a central dilemma that faces the electricity sector and other service sectors, which is the need to sustain ongoing service under changing circumstances while also making plans and investments that anticipate future demands.

LIKELY SCENARIOS

Based on their answers to the other supporting questions, we asked authors to present likely scenarios for storm-prone coastal areas. What unavoidable pressures will influence future actions, potentially improving or degrading resilience in coastal areas? Hazard experts generally support policies to encourage planned and gradual shifts of population and facilities away from the most dangerous areas along rivers, estuaries, and ocean shores, and to reduce the vulnerability of development that remains. What they envision actually happening is messier. Scenarios include long-term efforts to fortify properties and sites with high economic value, such as lower Manhattan or critical port facilities, protections for sites that have strong cultural or historic value, and gradual abandonment of development that is not economically viable. We discuss here projections included in several of the case studies for future storms and sea level rise, likely challenges to infrastructure, and the consequences for municipal revenues.

Decker and Robinson outline the problems of rising sea levels and the increased likelihood of severe storms that are projected by scientific computer modeling of climate change. Even storms entirely equivalent to past storms, when added to higher sea levels, will be able to send Sandy-size surge waters into coastal settlements. Demands from residents (Leckner and colleagues) and emergency managers (Hess and Conley), combined with federal fiscal needs for a less-subsidized flood insurance program, will increase political pressure over time for flood rate mapping to incorporate a realistic assessment of the potential impacts. However, the resulting rate increases for many will affect decisions regarding the affordability of at-risk locations. Weather forecasters themselves feel the need to sustain and improve their own technology and to increase international cooperation in forecasting to respond to these climate trends and to reduce harms.

Felder and Chandramowli's discussion of challenges for the electric utility industry in the United States has lessons for complex institutions of many kinds. There are long to-do lists for the electricity sector in the United States, which is burdened with aging infrastructure, highly complex and vulnerable physical and control systems, and changing generation sources. Although much of that sector is slow to change, its regular use of models and scenarios will help organizations in this sector identify emerging threats.

Andrews models the effects of disasters on local property taxes rates in several small towns and finds some surprising results. Property ownership in hazardous areas has been and continues to be subsidized by national policies in the United States. Allowing for owners of damaged residences to be bought out by federal or state programs would seem likely to lead to higher property tax rates for the remaining homeowners. His models show, however, that reduced costs of disaster response can actually be larger than the foregone taxes. Considering that broad political support for federal subsidies of flood insurance is declining, these findings could encourage local officials to support buyouts, or at least to be less supportive of redevelopment in the most hazardous areas.

LESSONS

- Overall, we have not yet seen transformative adaptations that are removing people from the most hazardous areas, in part because building barriers and planning for evacuation are still the key approaches used in countries that can afford them.

Over the past few decades, these two strategies have reduced the number of deaths from disasters worldwide (though Typhoon Haiyan/Yolanda is a stark contrary case), but because these strategies facilitate continued development, property damage costs have increased in many places. With a few exceptions, reactions to Sandy are enhancing these two trends in the storm's region. For the coupled strategy of barriers and evacuations to work well as climate threats intensify in the near- to middle-term, it will become even more important to incorporate disaster and evacuation planning into the broader cultures of coastal settlements, even in places with little hurricane experience, like the eastern portions of Cuba. Authorities rarely acknowledge that concentrating development in flood-prone zones places ever higher burdens on social arrangements such as evacuation planning. The fiscal consequences of rebuilding in these zones is also barely acknowledged by local officials, most of whom apparently remain convinced that development improves the fiscal outlook. Further, if a storm overcomes a physical barrier such as a sea wall, damages can be far higher than in the absence of the barrier, because landowners will be unprepared for the flooding.

- Some places are much more hazardous than others, and policies should address these zones first.

In Haiti, which had suffered an earthquake, a hurricane, and a cholera outbreak two years before Sandy hit, hazards became compounded. Hazards were also compounded in places like the highly polluted and surge-prone Gowanus Canal in Brooklyn, where hazards were created by affluence instead of poverty. Reducing vulnerable populations and sensitive infrastructure in highly hazardous coastal zones may be cost-effective even within a couple of decades. Adopting improved emergency preparedness and response measures targeted to these areas is cost-effective immediately and can involve governments and nongovernmental organizations.

- For social change to occur, advocates of change must overcome barriers. Changes may emerge both from the top down and from the bottom up.

Discussions in this book about settlement practices, market demand for land, perverse policies, exposed and aging infrastructure, and leaders' aversion to short-term spending for long-term benefits lead to pessimism about the likelihood that adaptation will occur quickly to significantly reduce hazards. Yet changes in public attitudes and in some organizations that were noted in several chapters suggest that methodical changes may spread. Concern about climate change and coastal hazards may seem diffuse, but if consumer, shareholder, or voter expectations change, leaders are likely to eventually respond with adaptive initiatives. Community-based institutions, such as fishing cooperatives and local groups in the Family Islands of the Bahamas, have also developed adaptive strategies that could influence others. Sharing of information and setting agreements for mutual assistance across jurisdictions and institutions like utilities is in itself part of adaptation.

- Ad hoc shifts are also likely over the next few decades, from those who cannot afford the costs of living or owning property in hazardous locations to those who can.

Many people are making a great deal of money from coastal property in some regions and find it worth making bets that they will continue to do so for some time. Interestingly, the increasingly wealthy owners of these properties are also likely to have more political capacity to demand policies that protect their investments, even while they comply with evolving policies to reduce damages through redesign, retrofitting, and rebuilding. Poor and even middle-class people, members of minority ethnic groups, the elderly, and people with handicaps are already more likely than others to be vulnerable to disasters, financially or physically. New forms of environmental injustice will arise if less affluent residents are essentially forced out and left to bear the costs of relocation on their own. A significant question is whether certain areas may be considered sufficiently risky or undesirable that wealthier people do not move in when the poorer owners abandon their damaged properties. If so, then market-driven abandonment may occur, irrespective of public policy.

- The success of adaptive strategies depends in part on our reading of the past.

It is understandable that people come to consider the long periods between storms as normal, treating extreme events as anomalies and misremembering actual impacts. Yet it is also possible to treat extreme events as part of the range of conditions that we should incorporate into our building plans and social practices. Adaptations would ideally be consistent with underlying natural processes, rather than attempting to counter those processes, because resiliencies in human and ecological communities are interrelated. Nostalgia for an imagined past can be unhelpful, although the history of human settlement does provide some examples of adaptive structures and practices.

- Strategies will have to change as climate change progresses.

Understandings of the past will have to be tempered by the realization that previous experience and local knowledge of hazards will become less useful as climate change advances. In some coastal areas around the world, recent sea level rise means that normal high tides are already regularly causing flooding. Climate change projections for the next fifty years indicate that problems like this will spread well beyond the lowest-lying sites, posing enormous challenges. Most designs and policies being offered today address only the near- to middle-term scenarios and will be outmoded within a few decades. Efforts to improve climate modeling and ongoing monitoring of the effects of climate change will be essential.

CONCLUSION

Hurricane Sandy was a major disaster for a few Caribbean islands and the third of three tropical storms that hit the mid-Atlantic region of the United States in two years, and by far the most damaging. Sandy may be most important regarding the questions it has raised about risks and responses, rather than the actual changes it causes to development patterns and policies. As more storms inevitably occur, these questions and the debates about them may well result in pulses of change that build toward a different future for at-risk areas.

Humility is in order, because while experts can suggest metrics for guiding rational decisions about climate adaptation, there may be no obviously optimal choice for some conditions. In addition, the concepts of risk and vulnerability are highly abstract. It is difficult to translate probabilities of extreme events into expectations about one's own safety or about buying a particular piece of property. The case studies here have described known risks and outlined expert suggestions, but the authors also acknowledge historical and cultural conditions that affect decision making. Some of the most important choices will necessarily be driven by what people value rather than by what experts suggest.

We end with an appreciation of the commercial fishermen interviewed by Oberg and colleagues. One said that the night the storm hit, he went out to check his boat and plan his next steps because "I'm a fisherman." Fishermen's preplanning, persistence, and adaptation in the face of risks seem useful lessons for others living on urbanized coasts. Yet there is an important difference. Fishermen must expose themselves to risks in order to pursue their livelihood, but many current coastal land uses do not have to be located in highly hazardous areas. How will other people adapt?

NOTES

1. Erin O'Neill, "Nearly 50 Homes Purchased through NJ Sandy Buyout Program to Be Demolished in Coming Weeks," *Star Ledger*, June 4, 2014; "Staten Island Homeowners

Offered Buyouts to Abandon Homes Hit by Sandy," *Daily News* (Reuters News Service), May 5, 2013; Matthew Schuerman, "Cuomo Expands Sandy Buyouts to 600 on Long Island," WNYC.org, 2013, http://www.wnyc.org/story/cuomo-expands-sandy-buyouts-600-long-island.

2. N. T. Brewer et al., "Risk Perceptions and Their Relation to Risk Behavior," *Annals of Behavioral Medicine* 27, no. 2 (2004): 125–130; Neil D. Weinstein and Mark Nicolich, "Correct and Incorrect Interpretations of Correlations between Risk Perceptions and Risk Behaviors," *Health Psychology* 12, no. 3 (1993): 235–245; Howard Leventhal, Kim Kelly, and Elaine A. Leventhal, "Population Risk, Actual Risk, Perceived Risk, and Cancer Control: A Discussion," *JNCI Monographs* 1999, no. 25 (1999): 81–85; M. Fishbein and I. Ajzen, "The Influence of Attitudes on Behavior," in *The Handbook of Attitudes*, ed. D. Albarracin, B. T. Johnson, and M. P. Zanna (Mahwah, NJ: Lawrence Erlbaum, 2005).

3. Terrence Dopp, "Sandy Becomes Another Christie Albatross," *Bloomberg Business*, June 5, 2014.

NOTES ON CONTRIBUTORS

CLINTON J. ANDREWS is a professor of urban planning at Rutgers, The State University of New Jersey, where he analyzes interactions between human behavior and buildings, infrastructures, and communities.

JOANNA BURGER is a Distinguished Professor of biology at Rutgers University who has conducted research for more than forty years in the three major bays of Delaware, New Jersey, and New York, concentrating on long-term studies of population dynamics, behavior, heavy metals, and conservation of birds within their ecosystems in a context of anthropogenic effects.

SHANKAR CHANDRAMOWLI, PhD, is an associate at ICF International, Inc., Fairfax, Virginia. Previously, he worked as a graduate assistant at the Center for Energy, Economic and Environmental Policy at Rutgers University.

PATRICIA M. CLAY is an anthropologist and researcher with NOAA Fisheries who studies fishermen, fishing communities, and the impacts of regulations and climate change on socioecological systems.

LISA L. COLBURN is an applied anthropologist and researcher with NOAA Fisheries who studies fishing community vulnerability and resilience related to the impacts of regulatory and climate change.

BRIAN W. CONLEY is a research analyst at the University at Buffalo, State University of New York, who has applied his expertise in geospatial analysis and urban sustainability to improve emergency and evacuation planning.

STEVEN G. DECKER, an assistant teaching professor in the meteorology program of the Department of Environmental Sciences at Rutgers University, is an expert in midlatitude weather systems who has participated in international weather forecasting contests while teaching students about synoptic and mesoscale meteorology.

FRANK A. FELDER is the director of the Center for Energy, Economic and Environmental Policy and a research professor at Rutgers University.

He conducts research on energy efficiency evaluation, renewable energy resources, wholesale and retail electricity markets, and the intersection of climate change and power systems.

JULIA A. FLAGG is a doctoral candidate in the Department of Sociology at Rutgers University, where she studies disasters, climate change, and political sociology.

KENNETH A. GOULD is a professor of sociology and the director of the Urban Sustainability Program at Brooklyn College, and a professor of sociology and earth and environmental sciences at the Graduate Center of the City University of New York. His research focuses on the political economy of environment, technology, and development, with an emphasis on the role of inequality in environmental conflicts and the implications of economic globalization for efforts to achieve sustainable development trajectories.

ROBERT B. GRAMLING (d. 2014) was an environmental and historical sociologist who studied coastal development in the United States and its social consequences, with a special focus on changes in lower Louisiana made by the oil and gas industries and by the federal government.

DANIEL BALDWIN HESS is an engineer and urban planner at the University at Buffalo, State University of New York. His research focuses on urban transportation systems and stressors placed on them, including extreme events that require large-scale evacuation.

MARK ALAN HEWITT is an architectural historian at Rutgers University specializing in research on vintage houses and a preservation architect with an active design practice.

BRIAVEL HOLCOMB is a geographer who, having circumnavigated the world twice as part of Semester at Sea, teaches a course on tourism planning at Rutgers University.

ASHLEY KONING is a PhD candidate in the Department of Political Science at Rutgers University and the assistant director of the Eagleton Center for Public Interest Polling.

MARIANA LECKNER, PhD, CFM, is an emergency management consultant experienced in federal, state, county, municipal, and nongovernmental programs in New Jersey, including Hurricane Sandy response and recovery projects.

TAMMY L. LEWIS is a professor at Brooklyn College and the City University of New York's Graduate Center who studies transnational social movements, urban sustainability, and sustainable development with a focus on Latin America.

BONNIE McCAY is an anthropologist and professor emerita of human ecology at Rutgers University who has researched fishing communities and climate change in the United States, Canada, and Mexico.

MELANIE McDERMOTT, PhD, has long studied community-based resource management in many contexts around the world, turning more recently to research pertaining to coastal resilience in New Jersey.

JAMES K. MITCHELL is a geographer at Rutgers University with more than forty years of experience studying human responses to natural hazards and disasters on four continents and advising public policy organizations.

LARRY NILES, a biologist with the Conserve Wildlife Foundation of New Jersey, has focused on population dynamics and conservation of migrant shorebirds and horseshoe crabs for more than twenty years and the restoration of Delaware Bay both before and after Sandy.

ANGELA OBERG is a doctoral student at the Bloustein School of Planning and Public Policy at Rutgers University who studies the politics of service delivery and human-environment relations in cities.

KAREN M. O'NEILL is a political and historical sociologist at Rutgers University who studies how policies on land, water, and hazards change the reach of government, the status of experts, and the well-being of various social groups.

DAVID P. REDLAWSK is professor of political science and the director of the Eagleton Center for Public Interest Polling at Rutgers University.

DAVID A. ROBINSON, a professor in the Department of Geography at Rutgers University, has served as the New Jersey State Climatologist since 1991 and in this role continually monitors weather conditions across the state and contributes information and insights to forecasting, emergency management, and media entities.

ADELLE THOMAS is a geographer at the College of the Bahamas who conducts research on the human dimensions of environmental change within the Caribbean, including intersections between disaster risk reduction and climate change adaptation.

DANIEL J. VAN ABS is associate professor of practice for water, society, and the environment in the School of Environmental and Biological Sciences at Rutgers University and has more than thirty years of experience in water resources planning and management with state government and nonprofit organizations.

INDEX

Page references followed by an f indicate a figure; those by a t, a table.